BABEL AND THE IVORY TOWER
THE SCHOLAR IN THE AGE OF SCIENCE

Babel and the Ivory Tower

The Scholar in the Age of Science

W. David Shaw

UNIVERSITY OF TORONTO PRESS
Toronto Buffalo London

ISBN 0-8020-7998-9

Library and Archives Canada Cataloguing in Publication

Shaw, W. David (William David)
Babel and the ivory tower : the scholar in the age of science /
W. David Shaw.

Includes bibliographical references and index.
ISBN 0-8020-7998-9

1. Learning and knowledge.   2. Knowledge, Theory of.   I. Title.

BD161.S43 2005   001.2   C2004-903173-2

University of Toronto Press acknowledges the financial assistance to
its publishing program of the Canada Council for the Arts and the
Ontario Arts Council.

University of Toronto Press acknowledges the financial support for
its publishing activities of the Government of Canada through the
Book Publishing Industry Development Program (BPIDP).

This book has been published with the help of a grant from the
Canadian Federation for the Humanities and Social Sciences,
through the Aid to Scholarly Publications Programme,
using funds provided by the Social Sciences and
Humanities Research Council

In memory of the mentors I still reach out to
'Tendebantque manus ripae ulterioris amore'

Contents

Acknowledgments

When I write on higher learning, I feel like a matador who steps into several bull rings at once, armed with neither a cape nor a dagger. Abetting me in this dangerous enterprise is a lifelong interest in poetry and science, which requires me to venture into areas of knowledge not my own. In an essay, 'Poetic Truth in a Scientific Age: The Victorian Perspective,' which I contributed to *Centre and Labyrinth: Essays in Honour of Northrop Frye* in 1983, I explored the Victorians' debt to the 'terrible Muses' of 'Astronomy and Geology,' as Tennyson calls them in his poem 'Parnassus' (l. 5). My interest in the two cultures gradually expanded to include the future of poetic education, and more generally the fate of the humanities themselves, in our own contemporary world of technology and science.

For his encouragement and counsel, I want to thank Lennart Husband, Humanities Editor of the University of Toronto Press. I am also grateful for the support of my wife Carol, for the sustained interest of my five children, and for the conversation and friendship of Eleanor Cook and Andrew Brack. A portion of this book's fourth chapter was originally given as a lecture, 'Babel and the Ivory Tower: Victorian Poetry in the Age of Science.' For an opportunity to present the lecture I am indebted to the organizers of a conference on the future of Victorian poetry held in March 2003 at the University of Western Ontario in honour of Thomas Collins and Donald Hair.

Twenty years ago in a chapter on 'education by poetry' in *The Lucid Veil*, I dimly discerned the shape the present study would take. But before I could explore theories of knowledge in higher learning I had to take a detour. Until I had examined traditions of learned ignorance and hidden knowledge in *Victorians and Mystery, Elegy and Paradox*, and

Origins of the Monologue, I could not grasp the distinction between two paths to knowledge. Nor could I understand how scholarship might challenge authority and renew tradition without aping science or trying to clone a Muse.

For the chance to go on this odyssey I am grateful to the Killam Foundation, which awarded me two Senior Research Fellowships; to the University of Toronto, which granted me a Connaught Research Fellowship; and to the Social Sciences and Humanities Research Council of Canada, which contributed a three-year research grant. Though it is harder to audit an intellectual quest than a bank account, I hope that (despite delays) the time and funds were not misspent.

Discovery and contemplation are as unique as Newton and Shake-speare. To talk with the great prophets, scholars, and poets of the past about justice, God, or Shakespeare's art is a much more erotic form of discourse than talking about quarks or genes, because (as George Grant has said) 'what is to be known about justice or God or beauty can only be known when they are loved.'

Babel and the Ivory Tower

In a technological society that assumes every phenomenon from the business cycle to dementia has an explanation or a cause, we are human-ized by things we cannot control or explain, like poetry or love. Just as a courageous inquirer lives the questions that have no answers, so a born scholar or critic is *not* drawn to his vocation for a reason. It is precisely because he is obsessed *beyond* reason with Plato or Mozart that he is justified in describing his career as a calling. What cannot be known in science, theology, or the choice of a vocation taps into mystery and keeps us human.

Babel and the Ivory Tower

Poetry is the breath and finer spirit of all knowledge; it is the impas-sioned expression which is in the countenance of all Science.

William Wordsworth, Preface to *Lyrical Ballads*, 1800

A house of scholars is never a mere seminary or trade school, but a phantasmagoria of shifting towers and caves. Often the house disguises itself as a casino in which enterprising gamblers try to make their intel-lectual fortunes in a competitive market. In its mad dance of disciplines, the gambler's cave exhausts the freakishness of arbitrary caprice. Some-times the murky labyrinth of the cave discloses the outline of an Ivory Tower, a home for reclusive scholars with the leisure to be wise. At other times the casino turns into an unsteady Tower of Babel, where energetic new thinkers compete to enter the conversation of mankind.

Babel and the Ivory Tower

An inquirer who seeks a single defining feature of higher learning – an attribute common to the maps and models of literature, history, philosophy, science, and professional education – seems at times to be lost in a maze without a plan. Sooner or later he experiences Yeats's alarm in 'The Second Coming': 'Things fall apart; the centre cannot hold; / Mere anarchy is loosed upon the world' (ll. 3–4). If pressed, however, to identify the secret of the maze, I should say that the Muse of higher learning – the scholar's master passion and the lifeblood of this mono-graph – is 'personal knowledge,' a phrase Michael Polanyi applies to the physical sciences but which I extend to every form of scholarship. A pro-fessor touched by the divine spark, who dwells on Parnassus with the Muses, appropriates knowledge as a living possession. As disinterested in principle as the objective knowledge of the new Prometheus, who is a robotic engineer called Frankenstein, 'personal knowledge' is the prize acquisition of both the old Prometheus and Spiderman. Though the Muse of higher learning cannot be cloned, she allows the prophet, rebel, and poet to soar, and the scholiast to weave intricate threadworks of commentary and gloss. Her energies of mind are the product of reflective inquiries which include the inquirer or view-holder as an essential feature of what is learned.

Babel and the Ivory Tower

Why should education be expected to 'take' in a society where the qual-ities of intelligence and wisdom are of necessity classified not even as by-products of its corporate life, but as waste-products? These qualities notoriously play no part in the production, acquisition and distribution of wealth, and therefore a social philosophy which regards this process as accounting for the whole content and purpose of mankind's exist-ence must write them off as so much slag.

Albert Jay Nock, *The Memoirs of a Superfluous Man*

It is an uneasy lot at best ... never to be fully possessed by the glory we behold, never to have our consciousness rapturously transformed into the vividness of a thought, the ardour of a passion, the energy of an action, but always to be scholarly and uninspired, ambitious and timid, scrupulous and dimsighted.

George Eliot, *Middlemarch*

For [Plato], truly (as he supposed the highest sort of knowledge must of necessity be) all knowledge was like knowing a *person*. The Dialogue itself, being as it is, the special creation of his literary art, becomes in his hands, and by his masterly conduct of it, like a single living person.

Walter Pater, *Plato and Platonism*

Whereas a scholarly Spiderman spins a web of argument or narrative, a prophet soars like an eagle. He is a Shelleyan Prometheus, whom Browning calls a 'Sun-treader,' communicating in flashes of aphorism and insight. Instead of weaving over time an intricate threadwork of argument in imitation of Francis Bacon's spiders, 'who make cobwebs out of their own substance,' ... the Sun-treader sees the world all at once as an undivided whole. In the tradition of Socrates and Jesus, he acquires knowledge not only through an exposure of logical fallacies in an adversary's judgments. He also acquires it through the more mysterious and incommunicable art of loving what he contemplates and embodying what he knows. Whereas the sciolist (or superficial pretender to knowledge) is impatient with all that is unsolved in his heart, the prophetic scholar heeds Rainer Maria Rilke's great injunction in *Letters to a Young Poet* to 'love the *questions themselves*': 'Do not look for the answers. They cannot now be given to you because you could not live them. It is a question of experiencing everything. At present you need to *live* the question.'

Babel and the Ivory Tower

BABEL AND THE IVORY TOWER

1

The Prophet and the Scholar:
Two Paths to Knowledge

The modern university is a great fortified post of information science and computers. As a house of learning, it is also a home for scholars who love what they contemplate and embody what they know. In a society in which book learning is an anomaly, the scholar must breach the citadel of computer wizards and technicians by combining his knowledge of books with the rebel's power to criticize authority, the prophet's power to renew tradition, and the poet's power to create a world that is no less true for being a vision. My portrait of the scholar has less in common with J.H. Newman's Oxford gentleman or T.H. Huxley's logic machine than with teachers like Socrates and Jesus, whose secret is their truths of embodiment, which they offer as alternatives to logical demonstration and rebuttal.

Robert Browning anticipates my distinction between prophecy and scholarship, wisdom and logic, in his elegy *La Saisiaz*, where he opposes the logical proof of a syllogism to the probation or proving-ground of life itself. Newman also insists that the living energy of the mind, its exercise of what he calls 'implicit reason,' has to be embodied in the whole life of a scholar, not just in his demonstration of a doctrine or a theory. Unlike W.B. Yeats's scholars, who 'cough in ink' and 'think what other people think' ('The Scholars,' ll. 7, 9), or Browning's pedant in 'A Grammarian's Funeral,' who 'decided not to Live but Know' (l. 139), a great prophet, rebel, or poet shakes the dust of archives from his feet and proves the truth by living it. As a poet in the German sense of *Dichter* (not an author of actual poems but a master spirit of literature), such a scholar holds learning and living, the wisdom of others and a knowledge unique to himself, in a delicate state of balance.

It has been said of Samuel Johnson that he was the only English poet on whom the Muse conferred a doctorate. It might be said with equal justice that the only doctor of letters graced by a Muse is the prophet or scholar who lives or embodies truths he can neither fully know nor demonstrate. In distributing her gifts, the Muse may confer both the bold surmise of the eagle-eyed Prometheus and the computer skills of his successor, a master of robotic engineering like Frankenstein. While an engineer is perfecting a technology and a scholar is designing an argument or weaving a threadwork of proof, a prophet is tracing an eagle's orbit round the sun. This book tries to relate the personal knowledge of the prophet to the scholar's knowledge, which is less hidden or mysterious, and usually more discursive. Can scholars and prophets transcend their points of view? Can they see the house of learning, not as a Babel of contending voices, and not as a casino for idlers who kill time with study, but as a new Athens, a city of art and culture, that keeps alive the soul of society as it passes from one generation to the next?

I finished writing this monograph on the morning I packed up the last of my books and left my college office for the last time. Quite literally, after four decades of teaching I was now without an academic home. But as a scholar in a technological culture, I sometimes think I was always homeless. In a world dazzled by the wizardry of Archimedes Silverpump in *Friendship's Garland*, a collection of satirical monologues by Matthew Arnold, the humanist who studies word play in Hopkins or irony in Browning may turn into an anachronism or ghost, an exile from Parnassus, without realizing it. Just as it takes light-years to discover that a distant star has died, so by the time the last card catalogue has been shredded and the last great library has been replaced by its website on the Internet, the world which once supported humanism may have slipped imperceptibly into history. Frankenstein frightens me, not because he invents a monster, but because his monster is a dangerously partial copy of its master. The new Prometheus is not a benefactor but an architect of technologies that displace what they clone. No one, I suspect, will quite remember when artificial intelligence replaced human understanding, when robots replaced people, or when the loss of personal knowledge, which was once the lifeblood of higher learning, became impossible to reverse.

My nightmare as a humanist is that I am a displaced person trying to transmit light from a burned-out galaxy. I am haunted by the fragility of culture and the shadow of lost knowledge. In a society that razes its monuments to scholars, I have no desire to raise a ghost. But I have no objection to conjuring the Muse of personal knowledge. As the genius of informed conversation and civilized life, the task of this Muse is to grace learning with intimacy and depth. Torn between the competing forces of Babel and the Ivory Tower, the singing school and the gambler's den, the modern scholar faces a dilemma. Should he try to preserve the soul of higher learning by cultivating the Muse of personal knowledge?[1] Or by imitating a technician who tries to clone the Muse, should he avoid the risk of being deemed a misfit? As an anatomy of knowledge in a society that models learning on science, this book is as much a eulogy as an elegy. It celebrates the scholar who proudly and justly claims that there are other ways of being intellectually open and imaginative than by copying a scientist or aping a technician.

The scholar's vocation, I take it, is a lifelong call to think adventurously and to live with risk as a prophet, critic, and architect of values. By 'critic' I mean a scholar who contemplates what he loves, and by 'prophet' a critic who embodies what he knows. Though scholars like Immanuel Kant and prophets like Friedrich Nietzsche take different paths to knowledge, their precision and daring should be complementary. As Kant himself might say, prophecy without scholarship is empty, and scholarship without prophecy is blind.[2] Unfortunately, in making science the new religion, our culture is constantly in danger of assimilating the scholar's first-person knowledge of the world to an unearthly dance of atoms that God himself would think twice before choreographing. When contemplating justice, God, or John Keats's odes, which only the intelligence of love can comprehend, a true scholar is an instinctive rebel. He refuses to sacrifice to the abstractions of the scientist, who uses concepts common to everyone, the personal knowledge – single, unrepeatable, compelling – that is unique to a self-conscious fashioner of worlds. Even empirical knowledge, which seems so solid and enduring, is as insubstantial as the hills in *In Memoriam* until the poet absorbs the scientific evidence into the substance of a personal vision that scrutinizes earth itself as a phantom shore.

> The hills are shadows, and they flow
> From form to form, and nothing stands;
> They melt like mist, the solid lands,
> Like clouds they shape themselves and go.
> *In Memoriam*, 123.5–8

A prophet who despises a cautious scholar may swell into an uncritical advocate of a political doctrine like Marxism or a cultural creed like the

New Historicism. By contrast, a scholar who has no prophetic ambition may collapse into a disenchanted positivist (a composer of chronicles, records, or literary histories that apply the scientific method to all branches of inquiry, including those to which it is inapplicable and alien). The best advice comes from Matthew Arnold's Empedocles: 'Be neither saint nor sophist-led, but be a man!' (*Empedocles on Etna*, 1.135). Instead of embodying what he knows by living the great questions, a zealous conscript is too often possessed by 'some wild Pallas from the brain / Of demons,' eager 'to burst / All barriers in her onward race / For power' (*In Memoriam*, 114.12–15). For this reason a dead Socrates is worth any living fanatic in the marketplace or forum.

At the opposite extreme, chasing the ghost of a departed faith in positivism, are the social scientists and funding agencies that ask scholars applying for a research grant to devise questionnaires, quantify their data, and include as many research associates as possible on their team. The best project is a megaproject, preferably one that drains away scarce resources from other scholars and takes a lifetime to complete. The protocols are so arcane that Canadian applicants need the advice of small research committees, consisting often of junior colleagues more attuned to current trends in cultural studies and social science than to issues in mainstream scholarship and criticism. As the product of a committee rather than a committed scholar, the grant application may fail. The senior scholar feels miserably jarred, of course, and his humiliation is public. But since all learning is assumed to approach the perfection of a soft science, his embarrassments tend to multiply until he shares Harold Bloom's frustration: 'I have lived to find the temples of learning consigned to amateur social work' (2002, 302). Presumably the ideal historian is H.T. Buckle, who contrives to make history a subspecies of statistics. Devoured by the vampire of technology and science, a hollow form with empty hands who feeds off a scholar's blood and accepts no substitute, the positivist believes that facts possess a mysterious power to organize and interpret themselves spontaneously. If scholarship is the ballast of higher learning and prophecy its sail, the cold touch of positivism is the torpedo that sinks the ship.

In an unguarded moment Keats admits that poetry 'is not so fine a thing as philosophy ... for the same reason that an eagle is not so fine a thing as a truth' (286). Though Keats's verdict is surprising for a poet who insists that even 'axioms in philosophy are not axioms until they are proved upon our pulses' (273), his reluctance to confuse an eagle with a truth, or a prophet's aphorisms with a scholar's proofs, tacitly assumes that there are two paths to knowledge. Whereas a scholarly Spiderman spins a web of argument or narrative, a prophet soars like an eagle. He is a Shelleyan Prometheus, whom Browning calls a 'Sun-treader,' communicating in flashes of aphorism and insight. Instead of weaving over time an intricate threadwork of argument in imitation of Francis Bacon's spiders, 'who make cobwebs out of their own substance' (Bacon, 514), the·Sun-treader sees the world all at once as an undivided whole. In the tradition of Socrates and Jesus, he acquires knowledge not only through an exposure of logical fallacies in an adversary's judgments. He also acquires it through the more mysterious and incommunicable art of loving what he contemplates and embodying what he knows. Whereas the sciolist (or superficial pretender to knowledge) is impatient with all that is unsolved in his heart, the prophetic scholar heeds Rainer Maria Rilke's great injunction in *Letters to a Young Poet* to 'love the *questions themselves*': 'Do not look for the answers. They cannot now be given to you because you could not live them. It is a question of experiencing everything. At present you need to *live* the question' (35).

At the heart of higher learning is an adventurous confidence in the power of enlightened minds to test, refine, and even transform the great models of scholarship and science. A community of scholars also celebrates more private and inward moments when the rightness of a theorem or a poem touches the mind then reaches into silence. 'There are only hints and guesses, / Hints followed by guesses; and the rest / Is prayer, observance, discipline, thought and action' (T.S. Eliot, 'The Dry Salvages,' 5.29–31).

An unfriendly reader might accuse me of making contradictory assertions about prophets and scholars. On the one hand, I align them as the common enemy of sophists and technicians. On the other hand, I set them apart by opposing their two paths to knowledge. My only defence is that the same contradiction surrounds the judicious scholar and witty poet in Alexander Pope's familiar aphorism: 'For wit and judgment often are at strife, / Though meant each other's aid, like man and wife' (*An Essay on Criticism*, 1.82–3). In practice, wit and judgment often sue for divorce. But in theory, they are a marriage of powers united in harmony, which honour and support each other. In Benedict Spinoza, for example, who builds like Euclid but towers like Moses, the architectonic power of the scholar and the oracular wit of the prophet come together in a single thinker.

Knowledge that makes an idol of scientific inquiry by banishing experienced quality from its third-person world treats the scholar either as a

superfluous person or as an unwelcome intruder. Such a culture is also
in danger of committing a form of the 'research fallacy' by reductively
assimilating to the physicist's discovery of natural laws the equally
demanding discipline of contemplating justice or of fondly but exactly
scrutinizing Keats's poems. To combat the fallacy of substituting research appropriate to subatomic physics for investigation into large
questions of social justice or art, this book formulates and explores six
theories of reflective inquiry, by which I mean knowledge that recognizes the view-holder as an essential feature of its world view (see p. 34).

Like the enabling medium of light, the power of knowledge is not
itself visible, though it is the source of all we see. To focus knowledge
and make its power visible, the university constructs a dome of many-
coloured glass. But like any idol, the dome is in danger of multiplying
variety in a wilderness of mirrors. Staining the light are three competing
tribes: merchants of Babel who sell knowledge as a marketable commodity; the ivory-tower dwellers who seek knowledge for its own sake; and
the Promethean benefactors who use knowledge to build a better social
community. Just as the self-serving sophist is a caricature of Socrates,
who makes knowledge its own reward, so Frankenstein, the father of
human cloning and robotic technology, is a mere parody of Prometheus, the prophet, deliverer, and healer of mankind. Unlike the
scholar in his ivory tower, who has a destiny but no destination, the
clamorous merchants of Babel value the goal more than the journey.
Instead of living the great questions like Socrates or Jesus, the sophist or
merchant of knowledge takes short-cuts to truth by selling his soul to
Mammon in return for profitable answers or useful short-term results.
For the open capacity of an explorer's experiments of living, sophists
substitute the tricks of a self-promoting lobbyist or hack. If their grim
descendants should ever establish ascendancy over the playful heirs of
Socrates, knowledge will cease to be personal: it will decline into a
dreary commitment to memory of rules and techniques.

My worst fear as a scholar is that the world I once thought I lived in no
longer exists. At some point the university where I taught, the intellectual home of Marshall McLuhan and Northrop Frye, has ceased to be
the Parnassus I remember, a dwelling place of the Muses where a culture was not so much mirrored as invented. The house of scholars that
was once a singing school for the soul is now a gamblers' den. Its new
patrons are less concerned with making their intellectual fortunes than
with winning the lottery of some high-paying trade like dentistry or
commerce. Whereas the scholar or prophet in his Ivory Tower is an

'intellectual Puritan in the narrowness of his affections,' the inhabitant of Babel, finding beauty in randomness and disarray, is more likely to be 'an intellectual Mormon in the [sheer] voluptuousness of his attachments' (Berlinksi, 206). Despite the efforts of some philosophers to see life steadily and see it whole, the view from Babel is too unstable to command assent. And even in staining the light with its many-coloured glass, the Ivory Tower excludes too much from its picture.

One original feature of my argument is its insistence that scientists, scholars, and professional practitioners must learn from each other. Mortimer Kadish suggests that unless professional education is studied in a liberal spirit, it is not professional. I go further than Kadish by arguing in chapters 3 to 6 that there are important heuristic, contemplative, and practical components in all knowledge – scientific, scholarly, and professional – that claims to possess the 'law-like' properties that make it teachable. In chapter 7 I show how personal knowledge is an antidote to the sophist's idolatry of 'know-how' techniques. Equally useful as a remedy for the technician's merely 'closed capacity' is the cultivation of 'open capacities' associated with the mastery of an academic model or paradigm, as I argue in chapter 8. The last three chapters explore the implications for higher learning of a second path to truth. As in Aristotle's theory of the moral virtues, political wisdom has to be studied in the practice and habits of statesmen who possess it. Like the 'implicit reason' of Newman's intellectual explorer (the prophet, rebel, or poet by whom truth is originally won), wisdom is not reducible to rules. No one knows better than Socrates that a logician's soul may be icy and Olympian. To know truth is also to become a friend of Socrates, who lives the questions he cannot answer and embodies the doctrines he cannot prove.

I locate traces of this second path to truth in the secret discipline of Donald Schön's coach or mentor, who imparts a half-intuitive knowledge or wisdom to 'reflective practitioners' of psychiatric counselling or architectural design. A charm to be used against the positivist's love of facts or the sophist's addiction to techniques of 'knowing-how' is the power of great scholars, prophets, and teachers from Socrates and Plato down to Blake and Nietzsche to seize intuitively the critical moment. Their grasp of revolutionary turning points in history transforms knowledge and allows them to cultivate disciples who reject apprenticeship (and even the name 'disciple') when their genius calls them. These prophets remind sophists of their culture's origins in lost wisdom and hidden powers. To appropriate such powers is to submit to a strenuous but secret discipline, usually imparted through the practice of a master.

Frederic Harrison, the positivist adversary of Matthew Arnold in *Culture and Anarchy*, predicts in his *Memoirs* that a technological culture's 'useless knowledge of useful facts' (Innis, 205) seriously impedes the creation of literature. As a cataract of factual information pours today over websites on the Internet, the deluge starts to 'asphyxiate the brain, dull beauty of thought, and' (in Harrison's words) 'chill genius into lethargic sleep' (2:324). The more relentlessly 'scientific and historical research piles up its huge record of facts,' its concordances, annotated editions, and copious bibliographies, the more worried even the positivist Harrison becomes that the 'scholiast's attention to minute scholarship' is accompanied by an 'inattention to impressive form.' He warns that as 'the printed book was the death of the cathedral,' so 'the school' today 'has been the death of literature' (2:327). An education in personal knowledge cannot revive the corpse of culture. But at least it may relieve the patient's symptoms and moderate to some degree the course of the disease.

For in a technological society that assumes every phenomenon from the business cycle to dementia has an explanation or a cause, we are humanized by things we cannot control or explain, like poetry or love. Just as a courageous inquirer lives the questions that have no answers, so a born scholar or critic is *not* drawn to his vocation for a reason. It is precisely because he is obsessed *beyond* reason with Plato or Mozart that he is justified in describing his career as a calling. What cannot be known in science, theology, or the choice of a vocation taps into mystery and keeps us human.

When the Muse of personal knowledge descends upon a scholar, his learning is tongued with fire beyond the pedantry of sophists and technicians. This is why a scholar-critic like Northrop Frye exemplifies the academy's boldest qualities. As a rebel like Socrates, Frye challenges the authority of unreflective custom and tradition. As a prophet like Prometheus, he brings a part of Olympus down to earth. And as a poetic visionary like Blake, he proclaims the essential sanity of genius and the madness of commonplace minds.

Having eagerly devoured Frye's books from the time I was in college, I never regretted my decision to apply for a post at Victoria College, where I could get to know him better. But he was also the colleague I talked to least. The conversation I now seek in his diaries and notebooks is a substitute for the conversations I never had during the twenty years we occupied adjoining offices. What a memento of vanity and waste, if one chooses to consider it – the books on Frye in the office next to Frye and the exchanges with the master I never had.

Universities invite scholars and scientists to tour Socrates' Athens or Yeats's Byzantium, the city of art and culture, and to study on their visit 'monuments' of the soul's 'magnificence.' But in order to impart a just appreciation of Byzantium, custodians of the holy city keep testing its visitors on the colour of the icons they have seen and on the intricate design of its mosaics. Whereas sophists promote the social utility and commercial value of the tour, Socrates believes that the intrinsic value of the knowledge to be acquired in the city repels with some authority the question: why visit Byzantium? When a professor at Memorial University in Newfoundland quips that 'we must humanize our scientists and simonize our humanists,' I fear that his simonized humanist, shining like chrome veneer, may also be a simonist who auctions off his portion of the Holy Spirit to the highest bidder. It is no doubt better to be Socrates than a sophist. But perhaps more worthy of imitation than either Socrates or the sophists are the Promethean deliverers who use the fire they steal from heaven to enhance the quality of civilized life. In using the light of knowledge to dispel the fog of ignorance and the shades of illusion in Plato's cave, a modern Prometheus may also want to explore a forgotten byway, climb a neglected tower, or even build a new temple in the city.

Since neither Athens nor Byzantium is a Benedictine monastery, its mission is to civilize the present and adorn the future by rightly celebrating the past: it should never try to humanize the world by caring more for a virtuous life than an intellectual one. Indeed it may seem incongruous for a university to serve as the conscience of society when many of its professors believe that the only authority to honour and the only value to preserve is the intrinsic merit of scholarly contemplation, scientific discovery, or personal knowledge for its own sake. But while it is preferable to be a temple of commerce than a tower of Babel, it is surely better to be Athens, the city of Plato's dialogues, than either. In the tower of Babel, Michael Oakeshott's 'conversation of mankind' is too often degraded into a babble of tongues. And in their imagined godhead, even the twin towers of a great commercial centre, built to last centuries, may become the short-lived target of hijackers and terrorists. In Babel we speak to our neighbours and deans; only in Athens, the city of culture, do we talk directly to posterity and God.

Like the church and the Bible, the power of all authentic education depends on wisdom and prophecy. To conserve tradition, it must honour Edmund Burke's primeval contract of eternal society, a bond between the living, the dead, and those as yet unborn. But by critically questioning and even subverting authority, a true education also risks breaks in continuity. As Frye says, 'wisdom culminates in such figures as Jesus and Socrates, who confront their societies with a transcendent vision which triumphs over those societies even as they reject it' (1988, 151). Since such a vision requires sustained contemplation, Socratic learning takes time. According to Frye, 'we must not worry about the time, but let [the methods] soak in. If rushed or directed by too active a will, the process collapses' (1988, 149).

An important function of all education is its initiation of citizens into the rites of the tribe. But Socrates also encourages each citizen to challenge the authority of custom or King Nomos by conducting experiments of living that are open-ended, risky, and unique to each inquirer. The tendency of any tribal mythology is centripetal. In the Middle Ages it builds up to a *summa* form, and in our own time it too often consists in surfing the Internet. According to Goethe, however, a culture which is not drawing on three thousand years of history is living from hand to mouth. When cultural amnesia threatens to strand society in a specious present, a function of education is to restore its memory. Though free enterprise in a liberal democracy encourages each learner to specialize, this fending for oneself is an affront to human dignity when it dissolves cultural memory, the glue that holds civilizations together.

Unfortunately, any recovery of one's cultural inheritance raises problems. As David Bromwich explains, 'a difficult paradox holds together the idea of a nonrestrictive tradition. Before it can be reformed intelligently, it must be known adequately; and yet, unless one recognizes first that it *can* be reformed, one will come to know it only as a matter of rote' (99). Even as historical education is striving to renew cultural traditions by raising the ghosts of the immortal dead, it must prevent such ghosts from acquiring false authority as a fourth person of the Trinity. Each discipline requires its great trailblazers, its Shakespeares or Einsteins, to transform tradition by defying authority and relaxing what Marjorie Garber calls 'the discipline of the disciplines' (83). Whereas 'the aims of any coherent and socializing culture' are purposeful and supervisory,' as Bromwich says (50), the scholar ought also to be that culture's most outspoken critic and adversary.

To examine higher learning in our own time is to examine, across a wide spectrum of disciplines, questions of research and methods of inquiry which differ so profoundly that Derek Bok believes the task may be 'beyond the capacity of almost any author' (1986, 3). Though I have no hope of succeeding where others have failed, I have drawn upon my experience as a scholar and upon a lifelong interest in academic models and paradigms to compare the contemplative vocation of the scholar with the scientist's passion for discovery and with the practical knowledge acquired in the great professional schools. In principle, all higher learning should be as liberal as philosophy or literary scholarship and as practical as an applied science like engineering or medicine. In practice, however, universities produce more technicians than scientists and scholars, more Edisons than Einsteins and Emersons. One reason why few graduates have any consuming interest in the life of the mind is that a utilitarian society leaves little room for the noble failure or the educated misfit. Its large suburban houses are not built for the scientist who is drawn to the cutting edge of non-commercial research or for the humanist who rises to the lofty, unwinnable challenges of philosophy or criticism.

I can tell when I have spent time in the company of a sophist, because I soon start having nightmares that a policeman will come into my lecture room and arrest me for impersonating a professor. In pretending to know more than he does, the sophist is to Socrates what a hypocrite is to a practitioner of irony. Though Socrates is the conscience of the academic mind, he is also its defender. He reminds us that deep down none of us really knows very much about our subjects. Fortunately, knowledge of this ignorance sometimes makes us wise.

The modern Socrates is genuinely forbearing. As an inquisitor of structures that resist analysis, he realizes that what is most worth knowing may not be teachable. Yet he has a genius for penetrating as deeply into the mystery of a moral or aesthetic universe as a physicist penetrates into the mysteries of subatomic particles and quarks. In an age when the humanities are in danger of being swept away by a rising tide of technology, Socrates affirms a connection between virtue and knowledge. Like the Renaissance humanists, he entertains the quaint belief that, by furnishing the scholar with instruments of power and understanding, such disciplines as philosophy, literature, and political science may still remain a nursery of wisdom, civility, and appropriate social principles.

Socrates' heir is a thinker of bold speculation, informed surmise, and the two-way thinking of a sceptic. He can reach as high as theology or as deep as physics. In his great dialogues, the art of discovery and the learner's response to cogent metaphor and civilized persuasion are an important part of what is learned. Whereas the sophist tries to reduce knowledge to a formula or set of rules, Socrates honours in the intelligence and mystery of wisdom an imagination that succeeds, an intuition that guesses some principle of experience. He is an energetic humanist who civilizes rational inquiry and uses love to urge the mind toward attainment. Unfortunately, the scholar's secret discipline and the prophet's efforts to struggle against the limits of what can be known and said discursively are precisely what a technically proficient sophist refuses to acknowledge. Mere group thinking sacrifices freedom. It conscripts the lucid moment of shared acceptance into commonplaces that are petrified and banal. In a world dominated by computers and machines, a technician tries to brush aside responses to beauty and humane feeling with an insulting smile or shrug of disdain, as if he had to educate idiots or children.

At the outset of his career the scholar is likely to be a self-critical Socrates, eager to test the limits of his understanding and his powers of invention. But competition for academic honours may turn Socrates into a compulsive achiever who seeks quick, spectacular results. Today's sophists are the academy's spokesmen for a new class of experts, computer engineers and technically sophisticated money managers and CEOs. John Guillory calls them 'free agents of pure charisma' (Fleishman, 114), superstars who are highly visible opinion-makers and sciolists but masters of nothing. As celebrities and gurus with no sense of vocation, the sciolists are 'smart' rather than scholarly, and informed rather than learned. Inseparable from the sophistry of these fallen scholars is 'the finish or polish' that Avrom Fleishman says 'extends to their dress and delivery, their sophistication in knowing who's who as well as what's what' (114).

One commentator, Allan Irving, believes that intellectual communities should reserve a place of honour for what he calls ironically the 'non-star' or 'good enough' professor. Being less compulsive than the 'star' performer who seeks research grants and endowed chairs, Irving's 'good-enough' professor is 'more attuned' to Socrates' willingness to remain in uncertainties. Prepared to master several disciplines instead of achieving eminence in only one, he also understands that no professor can be a specialist without being in the precise sense of the word an 'idiot,' an expert who knows more and more about less and less until he knows everything about nothing. The scholar who brings civility to his tutorials, curiosity to his lectures, and restlessness to his writings will be the opposite of a celebrity or star. As a wise genius of the place who has matured gracefully over time and who realizes with Aristotle that the best may be the enemy of the good, this generous mentor and sage is often the closest modern equivalent of Socrates.

Unfortunately, despite the pleasures of creative scholarship, the academic world is 'hierarchical and competitive; achievement is generally ephemeral and difficult to measure' (Getman, ix); and it is not always clear whether the modern professor is a recluse or a showman, a secular prophet or a technician, or some odd and improbable combination of Socrates and his adversaries.[3] Like George Grote, who argues in his *History of Greece* that Socrates was the greatest of the sophists, it is easy for the scholar in an age of sophists to mistake enemies for friends. Socrates is sometimes confused with his opposite, the sophist or technician who worships a servomechanism that lacks the life of reason and invention. Without the sceptic's chaste scruples, the sophist makes love, not to Pallas Athena, the goddess of wisdom, but to the many mechanical brides of a computer-based technology. In his shameless zeal to be useful, he is the most promiscuous drudge on God's earth.

As a democracy of talents committed to the dignity of difference, a university should cultivate, not the mindless conformity of a communist utopia, but an open and tolerant exchange of ideas, a kind of cultural laissez-faire. Regrettably, a democracy may also be defined as that form of government whose lowest shared desire allows Bentham to declare that pushpin is as good as poetry. If citizens in Bentham's pleasure-seeking democracy are equal before the law but unequal in talents, how are we to avoid what George Grant calls the tragedy of democracy, 'the gradual abdication by the higher faculties of their rule over man'? 'How,' Grant asks, 'does one reconcile one's deep loyalty to the tradition of democracy with the undoubted debasement of education that our democracy brings?' (1998, 177). Perhaps inspired and committed scholars, steeped in what they teach and study, can convert the Benthamites to the higher pleasures of rational conversation. Perhaps they can even give a few gifted students the courage to become teachers 'in a world which does not take the intellect seriously and which therefore thinks teaching is unimportant besides medicine or engineering, stock-broking or salesmanship' (Grant, 1998, 185). Though Grant tries to resolve the dilemma of educational democracy by appealing to the mystery of all people's being created equal, his final counsel is to live in the tension of the paradox itself. Instead of enforcing the lowest common denominator, the pluralism of a liberal democracy must try to combine the maximum economic equality with cultural diversity.

The modern academy is afflicted with a great deal of idolatry, or what Matthew Arnold calls machine worship – the mistaking of means for ends. The worst feature of idolatry is that (like academic politics or a fetishism centred on e-mail and computers) it distracts the scholar from the life of the mind. Perhaps the safest cure for idolatry is the diversity promoted by Michael Oakeshott's 'conversation of mankind.' In higher learning, as in culture at large, 'men require of their neighbors something sufficiently akin to be understood, something sufficiently different to provoke attention, and something great enough to command admiration' (A.N. Whitehead, quoted by T.S. Eliot, 1948, 50). Apart from Oakeshott's own bleak prophecies, there are few indictments of modern culture more scathing than T.S. Eliot's attack on education for failing to meet Whitehead's standards.

> For there is no doubt that in our headlong rush to educate everybody, we are lowering our standards, and more and more abandoning the study of those subjects by which the essentials of our culture – of that part of it which is transmissible by education – are transmitted; destroying our ancient edifices to make ready the ground upon which the barbarian nomads of the future will encamp in their mechanised caravans. 1948, 108

Anyone who is responsive to satiric animus and tone will recognize the similarity between Eliot's solemn banter and Oakeshott's spirited attack upon philistine culture.

In a prophecy that should be inscribed on the lectern of every lecture hall and carved over the portal of every college in the nation, Oakeshott predicts:

> A university will have ceased to exist when its learning has degenerated into what is now called research, when its teaching has become mere instruction and occupies the whole of an undergraduate's time, and when those who came to be taught come, not in search of their intellectual fortune but with a vitality so unroused or so exhausted that they wish only to be provided with a serviceable moral and intellectual outfit; when they come with no understanding of the manners of conversation but desire only a qualification for earning a living or a certificate to let them in on the exploitation of the world. 1989, 104

Though Oakeshott's tone is less shrill and solemn, the disaster he foresees harks back to the apocalyptic coda of Pope's great anti-epic, *The*

Dunciad, where the light of the human word is extinguished by the Word of uncreative dunces and sophists.

> Lo! thy dread Empire, *Chaos!* is restored;
> Light dies before thy uncreating word;
> Thy hand, great Anarch! lets the curtain fall,
> And universal Darkness buries All.
>
> *The Dunciad,* 4.653–6

A university that trains electricians and industrial managers instead of educating physicists and chemists has betrayed its mission. In certifying interpreters and cinema technicians instead of teaching language and literature, it has extinguished the light which orders our chaos and the power by which we live. Oakeshott's vision is as disturbing and difficult to ignore as his steady refutation of utilitarian logic, which like much technology and computer science today is, in T.S. Eliot's phrase, a 'great temple in Philistia.' As Eliot says of Arnold, Oakeshott has 'hacked at the ornaments and cast down the images' of this temple, 'and his best phrases remain for ever gibing and scolding in our memory' (1932a, 448).

The great battles of the present age are night battles, like the Syracusan disaster recounted by Thucydides and recalled by one of the darkest and most familiar wasteland poems of the nineteenth century.

> Ah, love, let us be true
> To one another! For the world, which seems
> To lie before us like a land of dreams
> So various, so beautiful, so new,
> Hath really neither joy, nor love, nor light,
> Nor certitude, nor peace, nor help for pain;
> And we are here as on a darkling plain
> Swept with confused alarms of struggle and flight,
> Where ignorant armies clash by night.
>
> Arnold, 'Dover Beach,' ll. 29–37

From nostalgia for his grand illusion, to an odd, reverberating, reproachful kind of lament, to the final surge of terror, all the steps of repressed despair, forcing its way slowly to consciousness, are embodied in Arnold's rhymes, grammar, line breaks, and metre. At first Arnold seems to be writing a love poem, and he keeps suppressing the dark side of truth. Only in the last verse paragraph does his nightmare vision break out unimpeded in a powerful echo of Newman's Oxford sermon on Faith and Reason. In Newman's own words, 'controversy, at least in this age, does not lie between the hosts of heaven, Michael and his Angels on the one side, and the powers of evil on the other; but it is a sort of night battle, where each fights for himself, and friend and foe stand together' (1887, 201). Between Babel and the Ivory Tower lies the darkling plain of Arnold's many exiles, 'wandering between two worlds, one dead, / The other powerless to be born' ('Stanzas from the Grande Chartreuse,' ll. 85–6). In 'Dover Beach' the incommunicable shadow of a lost knowledge – a unity of tongues – blurs into confused alarms of scholars and sophists, of friends and enemies of truth, whose clash is both 'superfluous and hopeless,' as Newman says in his Oxford sermon on Faith and Reason (1887, 201), because they fatally mistake each other in the night. The disaster in Thucydides, which both Newman and Arnold refer to, accurately foretells the appalling confusions in today's culture.

A book that tries to resolve the strains between the prophet and the scholar or between scientific and practical education by subordinating one to the other would be worthless. To pretend that the conflicts and risks confronting players in the lottery of academic life are not deep and often painful would also disqualify a writer from serious attention. But *Babel and the Ivory Tower* attempts, I hope, something less delusive. It identifies one possible solution to the great impasse between humanists and technicians in our culture by showing that prophecy and scholarship, wisdom and logic, are complementary. To this end I develop Ludwig Wittgenstein's distinction between showing or embodying truths, on the one hand, and logically demonstrating or proving them, on the other. Susanne Langer makes a similar distinction between the presentational forms of art and the discursive forms of logic and science. If poetry, as Shelley argues, is 'at once the centre and circumference of knowledge' ('A Defence of Poetry,' 517), then there is no reason why the presentational language of the cultural prophet, which is shared with the poet, should be sacrificed to the discursive language of the historical scholar, or vice versa. Each form of knowledge, though distinct, is united in the end in a single harmony, where each fulfils and enhances the other.

John Keats complains that 'with a puling infant's force,' Alexander Pope and his contemporaries 'swayed about upon a rocking horse, / And thought it Pegasus' ('Sleep and Poetry,' ll. 185–7). In his *Essay on*

Man Pope places Milton's Adam and Eve on just such a horse and thinks it wisdom.

> Created half to rise, and half to fall;
> Great lord of all things, yet a prey to all;
> Sole judge of Truth, in endless Error hurled:
> The glory, jest, and riddle of the world!
> *An Essay on Man*, 2.15–18

As a sophist who tries to reduce all knowledge to a technique, the scholar is the joke of our scientific culture. As a self-divided Adam who pursues a form of phantom knowledge divorced from the warm and breathing beauty his imagination finds delightful, the scholar is also a contradiction and a riddle. Finally, as a champion of personal knowledge – a kind of Prospero turning the shadow of science into the substance of vision – the scholar is not just the jest and riddle but also the abiding glory of a world in which he plays the part of Prometheus and our deliverer. The opposite of personal knowledge is not the mere falsehood or error of a failed scientist: it is Parnassus lost, the emptiness of a world from which all affective quality and value – what Newman calls 'the spontaneous pieties of the heart' – have been removed. Unlike the chaste austerity of science, whose rule is monastic in its rigour, the scholar's personal knowledge is comparable to Adam's dream of Eve in *Paradise Lost*, which is no less true for being imaginative and no less ravishing for being a vision.

As utopian portraits and satires of society and education, books like Arnold's *Culture and Anarchy* and Newman's *The Idea of a University* are unsurpassed. But their power does not diminish because they are only imaginary portraits or visions. *Babel and the Ivory Tower* is not a blueprint for educational reform. Nor does it aspire to the precision of a social science. In the tradition of Arnold and Newman, it tries instead to enlarge our stock of fresh ideas about the competing claims of maps and models, closed and opened capacities, in education. It also balances the claims of intellectual love and techniques of 'knowing-how,' of self-growth and utility, of wisdom and logic, in our culture. A university's claim to be Athens or Byzantium, the best of all possible cities, is usually only a libel upon possibility. In an eclectic academy every half-truth about bird migrations, Old Babylonian, or lunar politics may seem so true that any whole truth must be false. But if every idol in a holy place is a necessary evil, it must be good for a pilgrim to know the worst.

Freeing myself of traditional labels and taxonomies, I have allowed my reflections to move between an anatomy of learning and a memoir, a celebration of scholars and an elegy for their passing, always in the hope that what moves me will also interest a few select readers. Having recently retired from the profession, I have no incentive to burnish any idols or attack or defend a party line.

2

The Scholar's Wager: The Lottery of Higher Learning

Like F.H. Bradley's casino, the university is the one temple of God where 'worshippers prove their faith by their works, and in their destruction still trust in Him' (Bradley, 1930, aphorism 38). For the gambling analogy I am indebted to Mortimer R. Kadish, who compares the university to a lottery where every faculty member, student, and administrator is gambling his or her enlightened self-interest against the self-interest of everyone else. Kadish prefers the model of enlightened self-interest to a model of benevolent dictatorship, and even to the model of a democracy where the tyranny of a majority may still enslave an intelligent minority. The casino or lottery is an appropriate metaphor, Kadish thinks, because in higher learning there is no escape from risk. A monk who takes refuge in a singing school for the soul may repose with confidence on an axis of truth. But when a genuine scientist or scholar gambles on higher learning, he is putting at risk his most precious gift: he is wagering himself. In playing the game well, he may also be acquiring over time the profits that he seeks. For what he is risking is nothing less than the substance of his maturing energy and powers, the secret of a personal knowledge unique to himself. A gambler normally bets a small amount of money on expensive commodities like houses and cars. A scholar wagers a rare commodity – his personal knowledge – to increase the measure and value of the resource he is staking.

The expectations that education creates may not be satisfied. And even if they are, they may produce beautiful misfits or enlightened losers more finely attuned to the composed energies of scholarship or to the scientific search for exact truth than to the ready cut and thrust of the marketplace. A trained and dedicated mind may be ruined by his addiction. But the scholar's greatest risk is not to gamble at all. For should he refuse to stake his soul in a wager that repels with some authority the question why play the game, he will face the greater risk of forgoing the fitful light of the casino for the shadows of Plato's cave. And once a scholar or a scientist has been liberated by a wager, anything less bracing than the breath and finer spirit of personal knowledge is likely to be lethal.

As in John Rawls's blueprint for the ideal political state, the lottery of higher learning should maximize the satisfactions of all participants in the game before any player knows what his precise self-interest may be. A historian may have to decide whether to accept the small risk of failing at a conventional biography of Thomas Jefferson or the larger risk of failing at an ambitious study of the American Civil War. Should a lawyer try to prosper as a wealthy barrister known only to a small circle of colleagues and clients? Or should he try to speak to posterity as an Abraham Lincoln or Oliver Wendell Holmes? Even when the dice are rolled in the lottery of inherited intelligence and gifts, there are no guarantees. A world-renowned scientist may have a schizophrenic son. Illiterate parents may give birth to a scholar of genius. Like Kadish's metaphor of the lottery, Rawls's 'veil of ignorance' implies that the lottery of genetic inheritance and talents is the ultimate gamble. Paradoxically, however, 'the veil of ignorance' surrounding each roll, instead of increasing the uncertainty, makes the unavoidable bias or self-interest built into every institutional arrangement as disinterested and fair as possible. The idea is to minimize self-interest by agreeing upon the rules of the game before the results of any particular roll are known.

The casino's appeal is twofold. In the first place, toleration of different wagers is not just a democratic expedient but a virtue to be encouraged as a precondition of democracy itself. And secondly, though gambling is an inexact science, it has its own surprise rewards. Just as the notion of luck or genius is embraced as a challenge to the laws of probability, so in an otherwise uniform society pockets of indeterminacy are embraced as a welcome release from the tedium of what can be statistically predicted and controlled. Like the gamble of gifts, the lottery of higher learning is not a puzzle to solve but a mystery to baffle our intellect and keep us human.

In his great introduction to *The Republic*, which he uses as a blueprint for nineteenth-century Oxford education, Benjamin Jowett warns that 'the destination of most men is what Plato would call "the cave" for the whole of life ... There is no "schoolmaster abroad" who would tell them of their faults, or inspire them with the higher sense of duty, or with the ambition of a true success in life; no Socrates who will convict them of ignorance; no Christ, or follower of Christ, who will reprove them of sin ... The troubles of family, the business of making money, the demands of a profession destroy the elasticity of the mind' (2:161).

Like the casino of higher learning, Plato's cave is half-lie and half-art. But whereas the casino starts with fictions and ends with shared traditions and truths, the cave does the opposite. The casino's scribes lie to its citizens either to conceal a disturbing truth about its mummified theologies and myths or else to sustain a fiction that is civilizing and noble. By contrast, in sharing facts about their DVDs and computers, the cave people bewitch their minds with comforting half-truths or lies. A consumer society that worships its manufactured goods confers on its idols an illusory power and corruptive charm. In a surreal parody of the Eucharist, it literally devours the gods it worships. To remedy this distemper, the city and its academies, whether they be in Athens or Jerusalem, allow the shadows of the cave to die into the substance of the prophet's vision.

Built over a volcano on a flowery, habitable earth-rind, the city of art and culture is always in danger of being buried under lava. But a scholar's ultimate gamble is that as a potential prophet, rebel, or poet – a fashioner of new worlds and values – his mission is to save the city. Since each deliverer hopes to save culture in a different way, however, his precise mission is inscrutable. Even when the scholar's secret ambition is as exact as a science or as vividly realized as a poem, it is unique to each deliverer. And this unfathomable uniqueness, which is the life-blood of higher learning, is also the shoal on which most utilitarian theories of education founder. Theorists like Jeremy Bentham and James Mill, for example, insist that every gambler in the casino is entitled to equal treatment as a quantifiable magnitude, as a single shareholder in a stock company. But the gambler's identity as a prophet, rebel, or poet is of no account to these utilitarians. And neither are the gambler's loyalties or personal commitments to other players in the game.

To save society from the general shipwreck that results from a strictly utilitarian theory of value, John Rawls has argued in *A Theory of Justice* that an ethical system that tries to maximize the sum of all advantages may be less appropriate to an enlightened community of citizens or scholars than a contract doctrine of rights and duties. In universities, as in society at large, there are too few 'strong and lasting benevolent impulses' to prompt a faculty member or a student to endure 'loss for himself in order to bring about a greater net balance of satisfaction'

(14). By reading professors' manuscripts and offering extra seminars and tutorials to students who share Rawls's passion for political theory, the dedicated colleague and teacher may in the short run maximize the sum of intellectual benefits. But if Rawls's neglect of his own right to study and publish prevents his completing A *Theory of Justice*, does the net loss of satisfaction to future students of political science and ethics not ultimately invalidate his decision, even on utilitarian grounds? To be useful in the long run (at least in this case) is to recognize in Rawls's own words that 'the principle of utility is incompatible with the conception of social cooperation among equals for mutual advantage. It [even] appears to be inconsistent with the idea of reciprocity implicit in the notion of a well-ordered society' (14).

A strict utilitarian theory requires that all faculty members be treated equally. If a gifted scholar who has written influential monographs receives a large grant to write another book, should a senior colleague who has taught more classes but published nothing for thirty years also be rewarded for providing an admittedly different but (perhaps) equally useful service? Rawls would argue that such inequalities of treatment are just 'only if they result in compensating benefits for everyone, and in particular for the least advantaged members' of the community (15). The award of the prestigious grant to the productive philosopher cannot be justified, however, on 'the utilitarian premise that the hardships of some are justified by a greater good in the aggregate' (Rawls, 1971, 15). After all, who is to say that the philosophy department's conscientious Socrates has not offered more satisfaction to his students in the long run than the department's widely published Plato has brought to a small but select circle of scholars? Rawls would say that 'it may be expedient but it is not just' that the teacher 'should have less in order that [the published scholar] may prosper' (15).

An obvious way of justifying the inequality is to invoke a principle of rights and duties that runs counter to utilitarian doctrine. One can argue, for example, that in a research institution where one's primary duty is to advance knowledge rather than impart information, scholarship should be rewarded more handsomely than teaching. Even on practical grounds it may not be unjust to award a grant to the celebrated Plato if the situation of the less fortunate and obscure Socrates is improved as a result. Suppose the grants that the University of Western Ontario awards to its philosophy department depend on the number of research grants that the philosophy department itself receives from the Canada Council. Unless the local Plato receives his award from the Council, there will be no department fund from which to give the resident Socrates a teaching award. Socrates consents to the greater benefits now enjoyed by Plato because he knows that next year his own less-favoured position may be improved as a result. As Rawls explains, 'the intuitive idea' is that the privileged Plato can 'expect the willing cooperation' of Socrates and other less-favoured colleagues 'when some workable scheme is a necessary condition of the welfare of all' (15).

As in John Locke's theory of the social contract, the university's system of tenure decisions, promotions, awards, and merit pay must be understood as a situation that is purely hypothetical. 'Among the essential features of this situation,' Rawls says, 'is that no one knows his place' in the academic hierarchy, 'nor does any one know his fortune in the distribution of natural assets and abilities, his intelligence, strength, and the like' (12). Though higher learning is a lottery, its rules should be devised and ratified before players know whether the lot they have drawn in life has favoured them with the scholarly intellect of a Plato or the pedagogical gifts of a Socrates, or perhaps with some other gift. Otherwise, players may be accused of drawing up rules that favour the special talents of themselves or their offspring. No attempt to make the system fair, however, can disguise the fact that the game is still a lottery whose 'principles of justice' should be chosen, in Rawls's famous phrase, 'behind a veil of ignorance' (12).

Can people who have lost out in the lottery of life, or what Rawls calls 'the natural lottery in native assets' (104), be compensated in some other lottery? Though unequal gifts are simply a fact of nature and as such are neither just nor unjust, a principle of justice may require universities to allocate more resources to improve the long-term expectations of ethnic minorities, the rights of the gifted poor, and the special needs of the disabled. Graduate seminars on women's studies and black literature are not only permissible but also highly desirable if, by giving more attention to the appropriately qualified members of two minority groups, the seminar also improves the educational opportunities of some women students and some blacks. Otherwise, this concentration of resources on a favoured few may not be permissible. In Rawls's 'callous meritocratic society' (100), where rewards too often go to a clamorous elite of technicians and computer engineers, universities may also have to honour and reward the trailblazers and risk-takers among their scientists and scholars. Instead of trying to maximize the sum of advantages to all computer-literate and economically competitive members, universities may have to take measures to temper utility with justice and to balance short- and long-term social benefits.

Whereas Isaiah Berlin discerns a conflict between liberty and equality, John Rawls asserts that liberty has the prior and more important claim. Though Ronald Dworkin agrees with Rawls, it is only because liberty is a necessary precondition of equality, which Dworkin would otherwise rank ahead of liberty. Under a scheme of universal health insurance, the equal claim of all citizens to medical care assumes as its prior condition the free admission of a victim of severe chest pains to the emergency room of the nearest hospital. Without the right of prompt admission, the equal care of a nation's health system for all its citizens is meaningless. By analogy, we might argue that a prior condition of a university's equal concern for students wishing to study medical science or philosophy is the right of admission to a medical school or the freedom to become a philosophy professor. But no guarantees can be given, even to qualified and gifted students. For such guarantees are inconsistent with what Dworkin calls the 'model of challenge.' When the ambition of our life is to acquire the intellectual acumen of Dworkin or Rawls, then the denial to us of a tenure-stream appointment in Princeton's philosophy department is clearly a setback. But this misfortune or bad luck may also be a challenge to seek a post at Indiana or Toronto. We judge our life's success or failure, not by what Dworkin calls our 'volitional well-being,' a desire to live in the American Midwest rather than in central Canada, but by what he calls our 'critical well-being,' a lifelong ambition to teach and write philosophy, for example, when we fear that some alternative – the practice of medicine or law – would diminish us.

Equally important is the university's effort to provide each individual with a secure sense of worth. Only in this way will the lottery of higher learning begin to moderate the omnipresent risks. Though those who have won in the lottery of nature should utilize 'their good fortune,' the universities should also ensure that benefits are maximized 'only on terms that improve the situation of those who have lost out' (Rawls, 101). A university should assist its physically disabled students. But it should never try to change Emersons into Edisons or Mozarts into Einsteins. As Dworkin points out, we should not mistake the handicaps that prevent learners from meeting life's challenges with natural perimeters that define who they are. A schizophrenic Nobel laureate may have a 'beautiful mind.' Mental, psychological, and even sexual differences are an important part of a scholar's identity. A university has no mandate to interfere with these differences or try to change them.

One of the brightest polymaths I ever taught, who should have won every lottery in the casino, was too self-critical to be conventionally successful. He transferred from engineering school into an honours program in philology and literature, then later went from graduate studies in Anglo-Saxon and medieval Latin to medical school. He was 'our Sidney and our perfect man' (Yeats, 'In Memory of Major Robert Gregory,' l. 47); I loved his keen, ranging mind, his grace as a vocalist, and the surprising turns of phrase in his essays and letters. As a versatile humanist, his talents were Renaissance in their scope. Though he never mastered the credentials game, and though his expenditure of energy, mental and physical, was too intense, he had an immense capacity for self-formation and growth. I was touched by his tact, exuberance, and effortless authority in everything he turned his hand to. Like Dante's Brunetto Latini, he seemed not the loser among his degree-winning contemporaries but the winner in the only sense that matters.

Matthew Arnold quips that Socrates would be terribly ill at ease in Zion. He might be equally ill at ease in a lottery run by sophists and professionals. Too often we assume that the modern heir of Socrates is Isaiah Berlin, a gifted scholar and lecturer who was also an engaging conversationalist. But as Louis Menand observes, the most Socratic thinker in mid-nineteenth-century New England, the eccentric Chauncey Wright, was neither a professor nor a widely published scholar but a beautiful loser. A lonely genius who thrived on luminous moments of private conversation with a few committed students and friends, Wright was a founder of the Metaphysical Club, and a mentor to such luminaries as Oliver Wendell Holmes Jr, William James, and Charles Sanders Peirce. As a New England Socrates, physically uncouth but of great mental agility, Wright might have flourished in a small faculty seminar or in a tutorial for advanced philosophy students. Unfortunately, the Harvard that President Charles Eliot, a former chemistry professor at MIT, decided to transform into a modern university promoting science and professional study proved uncongenial to an inspiring mentor who seemed bland to the point of torpor in his public appearances and in the few scattered essays he wrote for learned journals. A casualty of Eliot's new academy, Wright had little charisma or classroom presence. His first course on psychology drew only twelve students. Years later, when he was only forty-five, Wright died suddenly after learning that President Eliot had declined to renew his course on mathematical physics because only one student had registered for it. The epitaph planted on Chauncey Wright by Louis Menand could aptly mark the grave of many Socratic exiles from today's academies: 'Wright was a modern intellectual in every respect but one: intimacy was the necessary condition for his mind to work. He was a type the university made obsolete' (2001a, 55).

When traditions is scholarship are changing as fast as fashions in clothing and cars, a professor's worst nightmare is an empty seminar room. Every fall I used to dread the moment when I had to ask the graduate secretary if any students had enrolled for my seminar. Facing failures of inheritance and denials of transmission, Wright's tragedy is like Zarathustra's or Nietzsche's own: he lacks disciples when he needs them most.

Like higher learning itself, an instructor's initial appointment to a college is more like buying a ticket in a lottery than investing in a government security. We underestimate the role of luck in most careers. My first offer of a post came from Victoria College, but it was written in a script I could not decipher. I have kept Kenneth Maclean's letter, and can still not read it. The second offer I never received. When A.S.P. Woodhouse at University College telephoned me at Harvard to ask why I had not responded to his letter, I had to tell him it had never reached me. It had gone to Oreste Rudzik in Chicago: apparently Oreste's own offer had been lost somewhere in Cambridge. By the time these mysteries were unravelled I had accepted a lectureship at Cornell, where I spent seven years in the wilderness of upstate New York before returning to Toronto as a prodigal son.

Since universities both shape and mirror society at large, I have thought it worthwhile in this chapter to search for political models or analogies of a successfully functioning university, which David Damrosch characterizes as a 'blend of engagement and withdrawal' (37). I prefer Mortimer Kadish's model of enlightened self-interest to a democratic model that tends to devise a tyranny of the majority, or to a 'tempered despotism' that, in Claude Bissell's phrase, has the virtue at least of 'speed and despatch' (1974, 121). But because the ugly genie of envy can destroy the soul of even the most carefully designed system, there is no guarantee that any single academic model can be successfully applied. Whereas David Bromwich opts for an intellectual community of one, Damrosch favours more scholarly collaboration: 'Either fellowship or death' is his final counsel (1995, 214).

Instead of offering a host of rewards ranging from religious salvation at one end of the spectrum to membership in an elite social class at the other, the university should candidly acknowledge that its lottery can as easily bankrupt its patrons as make their intellectual fortune. Just as a lover may be clumsier in his technique than a skilful seducer, so a student who cares deeply for science may be clumsier or less adept at it than a mathematically gifted student who shows only disdain for laboratory research. 'Love for a subject is logically independent of being good at it' (Passmore, 1980, 193). Such is the painful truth of trying to make one's fortune in the casino of higher learning. 'The will's somewhat – somewhat, too, the power – / And thus we half-men struggle' (Browning, 'Andrea del Sarto,' ll. 139–40).

Nonetheless, in trying to justify his claim that 'the task of reconciling effectiveness and the interest in self now constitutes the distinctive problem' of 'higher learning' (28), Mortimer Kadish gives six reasons why a studious and ambitious youth in search of his intellectual fortune should still place a wager on personal knowledge. In the first place, risk is unavoidable. Even if one avoids universities altogether or opts for the salvation of a seminary education or the higher vocationalism of dentistry or journalism, one still runs a double risk. Either one saves one's soul by becoming irrelevant or one becomes relevant by losing one's soul. As Kadish observes, 'to avoid facing a risk is not the same as avoiding the risk' (41).

In the second place, to gamble in the casino of higher learning is to expand the range of social and intellectual games one is qualified to play. It also enhances one's self-esteem by 'living a life worthy to be called human,' as Matthew Arnold says, and according to 'the power of intellect and science, the power of beauty, the power of social life and manners, – these are what Greece so felt, and fixed, and may stand for. They are great elements in our humanisation' ('Equality,' 1961, 484). Even though the powers of knowledge, beauty, and conduct may be in conflict with each other, Kadish argues that 'gaining freedom in one's estimation of oneself is a substantial reason to assume the risks' of becoming a self-divided rather than a whole person and of 'fouling up' (41) as a result.

The 'humanisation' of man in society is a third reason for making a wager on personal knowledge. Like Yeats, we accept the risk of sailing to Byzantium, the city of art and culture, even though our citizenship there may drain us of material wealth and make us in the end sadder and wiser people. A student of mine, who planned to enter medical school and join his father in the practice of medicine, fell in love instead with philosophy and literature. Though he was a brilliant undergraduate and a fine graduate student, he failed to find an academic post worthy of his talents. I am still convinced that a literary education was exactly what he needed to develop his special gifts. But his education in personal knowledge impaired his ability to earn a living. The only experience more distressing than getting less than we bargain for in liberal education is the experience of getting more, for by converting a business major to the life of the mind it may unfit him for commerce. To lack education may be a source of discontent. But it is by no means obvious that either personal or professional knowledge makes one happier. With alarming clarity George Eliot's Casaubon dramatizes the danger of receiving an education for which one is intellectually disqualified. 'To be trained, taught or instructed,' as T.S. Eliot warns, 'above the level of one's abilities and strength may be disastrous; for education is a strain, and can impose greater burdens upon a mind than that mind can bear. Too much education, like too little education, can produce unhappiness' (1948, 100).

A fourth reason for gambling on higher learning is that the faculty who run the games are on the same side as the players. Since each group wants self-growth and effectiveness to cohere as much as possible, at critical moments in the game their 'positional interests' may start to coincide. When Aidan Johnson sits outside my office armed with questions, refusing to leave until I agree to give him a year of tutorial instruction on the theology of Oscar Wilde, he is doing what every true student ought to do: he is making an unmitigated pest of himself. And a professor who turns away such pests is unworthy of his calling. Before young scholars can acquire the heightened power of practical decision that comes with a decline of interest in merely 'thinking for themselves,' they may first have to infect some vulnerable professor with a passion for Wilde's eccentric theology or odd aesthetic theories. If we reorganize the university to ensure the greatest utility and efficiency of its faculty, what happens to the unconventional scholar who swerves away from his own fixed schedules to tutor a wayward genius? If we opt instead for each member's pursuit of his individual whim or bliss, the university may lose its temporal authority and centre. Either way the inquirer seems to be trapped in a paradox or dilemma. 'First, the majority is always right,' as Northrop Frye says, 'for the majority is the source of temporal authority. Second, the majority is always wrong, for it is not the source of spiritual authority' (1964, 313).

A fifth reason for playing the lottery is that, despite the odds against it, there is always a small chance that, as a result of some anomalies in a discipline's paradigms or of some shift in its models, a genius like Galileo or Newton will unexpectedly emerge to revolutionize our under-standing of astronomy or physics. By disturbing the conventions or inventing new paradigms, trailblazers may suddenly see a new planet or galaxy swim into their ken. Kadish believes 'it is rational' for the adven-turous artist, scholar, or scientist 'to incur serious risks' in pursuit of ends whose rewards may be 'enormous' (42).

A sixth reason for entering the casino is that, like Hermann Hesse's Glass Bead Game, the pastime itself is immensely rewarding. Taking a chance on higher learning is like listening to Mozart: in James Cam-eron's phrase, it 'repel[s] the question *why* with some authority' (14). To play at all 'seems much more like exercising a right than fulfilling a duty' (Cameron, 15). More alluring than the prospect of succeeding in a profession or even growing as a person is what Kadish calls the 'adven-ture, wonder, fulfillment, novelty, fraternity, competition' of higher edu-cation and the sheer pleasure of the experience itself (42).

Since knowledge unique to each explorer is the lifeblood of higher learning, the first and most important way in which scholars can pro-mote this knowledge is by substituting for scientific experiments what J.S. Mill felicitously calls 'experiments of living.' Since no values are immutable or valid for all time, and since the humanities are not just a rudimentary form of science, Mill thinks that scholars have to explore all sides of the truth by giving 'free scope ... to varieties of character' and by discovering how 'the worth of different modes of life' can 'be proved practically' (J.S. Mill, *On Liberty*, 1947, 56). To cover the full range of options these experiments should be conducted, not in a scientific labo-ratory, but in the *theatrum mundi* William James constructs in such books as *The Principles of Psychology* and *The Varieties of Religious Experience*. Though an experiment of living can be playful, competitive, or purpo-sive, play can easily be profaned as frivolous triviality; games can degen-erate into mere contention or fractious dispute; and even a socially useful purpose can be degraded into mere technical training devoid of guiding principles or practical methods.

A second way in which scholars can promote personal knowledge and self-growth is to imitate the utilitarians and even the disciples of Charles Darwin by making what *is* a measure of what *ought* to be rather than the other way round. As Samuel Johnson tells Joshua Reynolds, the scholar seeks release from 'the *disease* of his mind' by accepting the 'cleansing challenge of objective fact' (Bate, 122). Because ignorance of their igno-rance is the worst malady of the untutored, scholarship must always be an exercise in what Nicholas of Cusa calls '*docta ignorantia*' or the ignorance of the learned. As one commentator says, 'without this agnosticism humans tend to move to the great lie that evil is good and good evil. In Christian language this great lie is to say that providence is scrutable' (Grant, 1998, 480). Most scholars operate on the assumption that there is a way things are, but that only a fool or a sophist presumes it is ultimately knowable or known. Scientists make a comparable assumption when they posit as a first postulate of faith the uniformity of nature. They do not know that the world is intelligible. But until they make this assumption, they cannot affirm the value of seeking knowledge, not because it can be converted into some other desirable commodity, like the jackpot in a poker game, but because it is desirable as an end in itself, like virtue or good health.

A third mark of personal knowledge is its education in eros or love, which urges the soul toward attainment (as Plato realizes) and helps Socrates glimpse a daily beauty in the lives of his friends. But to cure the distemper of Pygmalion, who fell in love with his own sculpture, the scholar must substitute for a mere projection of the idols of his cave a resourceful power to discover and invent. As I show in chapter 7, a model's open knowledge is to the closed knowledge of a map what Isaiah Berlin's idea of 'negative liberty' is to 'positive liberty.' Berlin believes that a liberal democracy's most valuable legacy is not freedom *for* but freedom *from*. What confers dignity is not the duty of a free citizen to save the world *for* democracy, the war against terrorism or any other political crusade. More important than positive liberty is the negative freedom *from* mental coercion and tyranny, including the tyranny of democratic majorities and their roadmaps to Utopia. Pursuing the same line of reasoning, Bertrand Russell argues that the most valuable part of any education is not the information that the memory stores passively or the positive knowledge that enslaves the mind to protocols and rules. Russell fears that 'compulsion in education ... destroys originality and intellectual interest' (215). What truly endures is the mastery of models that gifted apprentices are free to teach themselves once they have acquired the necessary information and internalized the first axioms or rules of their discipline.

Though freedom to appropriate the models of literary criticism and philosophy is a product of rigorous study, the duty of universities to provide such freedom does not then authorize them to indoctrinate the thinkers whom they free. Unlike soldiers in an army, scholars cannot be recruited into the ranks of some militant ideology promoted by a propaganda office of the state. In other words, the necessary negative liberty should not be confused with positive liberty. A scholar who has been liberated by a genuine intellectual model is unlikely to submit to the yoke of a mere roadmap or charter. New dogmas may be promulgated from Allan Bloom's pulpit at the University of Chicago, from the Yale School of Criticism, or from the latest citadel of Marxist scholarship at Duke. But such dogmas are more likely to resemble a commandment from Moses or an encyclical from Rome than the supple arguments of a truly liberated thinker like Socrates or Jesus.

A final attribute of personal knowledge is a scholar's willingness to criticize authority and renew tradition. Instead of producing mental slaves, universities should cultivate habits of mind that are as critical as they are constructive. Scholars must be free to challenge the disciplines that fathered them and even revolutionize them from within. As J.S. Mill realizes, the desire to be inert by insulating knowledge against hostile criticism is a sick desire. Without losing integrity and self-respect, free enterprise in the realm of ideas tries to maximize rather than minimize dissent. As David Bromwich says, 'you agree to tolerance not out of kindness to the claims of the other person but from an irreducible respect for yourself' (162).

Quoting Erwin Panofsky, Bromwich defines a Socratic humanist as someone 'who rejects authority' but 'respects tradition' (101). A sophist does the opposite: in paying homage to some pedagogical or professional authority, he resists any inquiry that might produce an unwelcome verdict. Unchallenged authority is a vestal virgin. Unless she is wooed and made love to, there will be no progeny to nourish, no legacy to preserve, no living tradition to venerate and transmit. When scholars reject authority, they should also try to preserve and renew what is most worth saving from the past. 'A people that no longer remembers,' Aleksandr Solzhenitsyn warns, 'has lost its history and its soul.'

In suppressing a conviction that the scholar's singularity is an important part of what he knows, the assumption that all knowledge aspires to the condition of inductive science drains away the lifeblood, breath, and fir.er spirit of learning: as Karl Popper argues, it may even impoverish science itself. For a forgotten secret of higher learning is that, while a scientist is proposing and testing theories about the physical world, a scholar is proving his truths by living them. The most important wager he can make is the gift of himself. Since scholars love what they contemplate but profess mainly what is dead in their heart, they often embody rather than speak their deepest truths.

In a society that worships science as the only respectable model of inquiry, there is a danger of conferring too much prestige on observation and induction. As Victorian critics of J.S. Mill's *System of Logic* have argued (especially F.H. Bradley in his *Principles of Logic*), it is no dishonour to ethics, religion, or poetry that they should show or exhibit what they cannot formulate explicitly in a descriptive proposition. When tongued with flame, intimate communications are shown rather than stated. They flare up in moments of sudden illumination, in the power of people like Mrs Moore and Aziz in *A Passage to India* to love or befriend each other across great divides of race, cultural difference, and age. The aura of a meaning we cannot lay hands on is familiar to every admirer of music and the visual arts. We also find it in the sublime parables of the Gospels and in the aphorisms of Nietzsche and Schopenhauer, whose truths must be perceived instantaneously and as a whole or not at all.

To appraise the two paths to knowledge (one the orbit of a sun-treading prophet; the other the threadwork of a discursive scholar), this book attempts what no other study has tried to do: it examines the power of both rational inquiry and embodied wisdom to criticize authority and renew tradition across a wide spectrum of disciplines. Like Arnold's saving remnant in *Culture and Anarchy*, who make their humanity rather than their class trait prevail, true scholars combine respect for tradition with deep suspicion of unquestioned authority. As the energetic civilizers and unacknowledged legislators of our world, they cultivate knowledge that is scientific rather than technical (chapter 3); wide-ranging and reflective rather than minutely specialized and partisan (chapters 4 and 5); applied, lived, and tested rather than dogmatic and untried (chapter 6); committed rather than impersonal (chapter 7); open rather than closed (chapter 8); and unconscious rather than preconceived or planned (chapters 9 and 10). We should never mistake the casino of higher learning for a bank, a gambler for a prophet, or the lottery itself for a deposit of certainty and truth. Like Darwin's theory of accidental variation, education ought to be a battle against the bewitchment of the mind by sciolists who underestimate the share of conjecture and uncertainty in what we know.[1] This battle is not an arid academic conflict but a life-and-death issue. For the future complexion of our culture is at stake.

3

The Scientist's Knowledge:
The Genius of Discovery

One axiom of this inquiry is that modern science has more in common with great scholarship than with technology. Both science and humane scholarship depend for their breakthroughs on the disciplined imagination and the adventurous thinking of highly original investigators. As one commentator says, 'the sciences and their histories are thrilling in similar ways to the arts and their histories, and the two interconnect on multiple levels' (Magee, 132). Even when science seems to build on a secure foundation of knowledge, it is often a blind date with such capriciously complex phenomena as next week's weather. Moreover, like a new school of economics or history, a scientific paradigm seldom explains all the facts. Whereas a geometrical proof logically coerces our assent, we accept a new paradigm in science with the free assent or informed faith we accord any reasonable belief. Individual complacency is the death of any scientist or scholar. Collective complacency is the ruin of universities. It is doubt, restlessness, anxiety, agony of mind – above all, a steady and luminous contemplation of what is still unknown – that nourishes discovery and growth in any discipline.

The house of intellect is divided into many mansions. But the next three chapters argue that unless the professions are studied in the liberal spirit of the scholar and the scientist, they will not be professional. And unless the scientist cultivates the contemplative and speculative habits of the scholar, he will be ill equipped to replace his models when the next scientific revolution requires a new paradigm to be invented. Even the scholar who fails to cultivate the informed habits and concrete speculations of the scientist will succumb to what Whitehead calls the 'second-handedness of the learned world,' which 'is the secret of its mediocrity. It is tame because it has never been scared by facts. The main importance of Francis Bacon's influence does not lie in any particular theory of inductive reasoning which he happened to express, but in the revolt against second-handed information of which he was a leader' (1949, 61). The opposition between a technical education and a liberal education is false, because 'there can be no adequate technical education which is not liberal, and no liberal education which is not technical: that is, no education which does not impart both technique and intellectual vision' (Whitehead, 1949, 58). Instead of contemplating the law of gravitation in a vacuum, Newton immediately calculates the attraction of the earth on an apple. A literary scholar who ascends from particular sonnets by Shakespeare and Milton to a general model of the sonnet will find his induction is barren unless, like one of Jacob's angels, he descends again to a particular sonnet by Keats to show how its fusion of techniques used by Shakespeare and Milton is resourceful and alive.

In examining the scientific, contemplative, and practical components of knowledge, I argue that all higher learning should ideally combine the genius for discovery associated with science, the contemplative power of humane scholarship, and the practical power that is the most important contribution of professional education. Scholarship is to science what the language games of Wittgenstein's *Philosophical Investigations* are to his mirror of pictured facts in the *Tractatus.* The third category of higher learning, professional education, consists of intellectual habits and skills that can neither be taught in a lecture hall nor learned in a library. They are best acquired by an apprentice in close contact with a master of her craft. Practical knowledge is the opposite of technical knowledge, though the two are often wrongly equated. Whereas a computer program can be designed to produce a Bell curve of grades with predictable results, the conduct of original science or scholarship is a far riskier activity whose outcome is never known in advance and whose success is never guaranteed. Each requires an artistry or talent that cannot be taught, a form of judgment that allows a learner to know intuitively when her scholarly models or scientific paradigms are leading her astray.

Practical as opposed to technical knowledge tells the chess-player what series of moves he should make. With a flash of insight it allows the experienced scholar to distinguish the profitable line of inquiry from the cul-de-sac. Even the knowledge and intuition that an experienced scientist possesses are not to be found in a handbook of ready-made formulas and rules. Nor are they to be sought in a magical talisman called the scientific method, a charmed password or key to open all doors. As

Michael Oakeshott concludes, the tyranny of a rational method or technique 'is the work, not of the genuine scientist as such, but of the scientist who is a Rationalist in spite of science' (1991, 35).

The great lesson of higher learning is trade-off and exchange. Thomas S. Kuhn and Michael Polanyi both discuss the special circumstances that make research in the physical sciences uniquely successful. But on the altar of its proven methods science also makes some harsh sacrifices. Most importantly, it sacrifices the humanist's first-person picture of the world. And in order to concentrate its energy by focusing on current research, science minimizes history and tradition. Scientists are encouraged to refine and supersede the models of Newton and even Einstein in ways that no literary scholar can ever forget the achievements of Dante and Milton. As Mortimer Kadish explains, 'people in the sciences are less concerned than those in the humanities with building Memorials to past greatness ... The sciences are present- and action-oriented. People do science; ... Few, except for philosophers of science and culture, now need to read Newton as writers, critics, and dramatists need to read Shakespeare' (100).

Though Whitehead says 'a science that hesitates to forget its founders is lost' (quoted by Kuhn, 137), a culture without courageous founders has no tradition to preserve. Unless science finds some means of remembering its heroes while revising their theories, it will suffer the worst effects of amnesia. To live in a specious present is to experience the cultural equivalent of Alzheimer's disease. A society of ant-like technicians is a society of Struldbrugs with memories too frail to carry them from the beginning of an experiment to its end. Every scholar has to honour the god Janus, who looks backward and forward simultaneously. But a scientific culture has tunnel vision, which it directs exclusively toward the future. A culture devoted to the scholarship of discovery may have to invent new ways of stepping off the moving platform of the present if a potential asset is not also to become its greatest liability.

There is the further problem that, though a scientist's discoveries are the paradigm par excellence of pure research, they can seldom be duplicated outside science itself. And since science's success depends on developing often contradictory talents for innovation and preservation, rupture and continuity, there is no guarantee that it can produce a trailblazer like Newton or Einstein when the time arrives for another revolution. Indeed the conscientious application and predictable development of a scientific model may destroy the precious passion for novelty and invention that makes possible the discovery of new paradigms. Ironically, the more victories science achieves, the more it is in danger of abolishing itself. Not in the trite sense that computers will answer all its questions, but in the more troubling sense that its technology will substitute a slave mentality for the intelligence capable of mastering nature and finding new theories and solutions.

Because it requires an unusual combination of qualities, a scientific vocation is rarer than most people think. Ideally, a scientist unites a computer's capacity to function as a clear cold logic machine with a scholar's commitment to an austere intellectual discipline, which is at once passionate and rigorous, intense and controlled. A scientist should exhibit and value that 'chastity of the intellect' which George Santayana praises as the most valuable legacy of a philosopher's impulse to test ideas and criticize authority. Unlike most philosophers, however, a scientist can afford to be relatively untroubled by what Allan Bloom calls 'the great theoretical difficulty of modern science – that it cannot explain why it is good' (1987, 50).

A scientist may develop a materialist hypothesis of the physical world, not because it is more or less true than an idealist hypothesis, but because it has practical utility. T.H. Huxley commends the symbolic fictions of the physicist for their experimental economy and fertility. As Richard Rorty argues, a scientist normally tells a story of a stable physical world against a background of a changing organic world, not because it is necessarily more accurate than the alternative, but because (at least in Western cultures) it creates a more credible narrative of an entity the mind can mirror. A science or philosophy with a practical turn of mind will choose as its epistemological model the mirror of nature rather than the lamp of a mind furnishing essential ingredients of its own knowledge, even though the lamp more accurately mirrors what is happening.

The scholarship of discovery may also allow a scientist with a contemplative turn of mind to pursue forms of so-called 'pure research' that develop innovative explanations in astronomy or physics. A scientist's heuristic passion must be creative in the sense Michael Polanyi has in mind when summarizing Kepler's discovery of the five solids among the heavenly orbits. 'We call [such] work creative because it changes the world as we see it, by deepening our understanding of it' (Polanyi, 143). As Polanyi reminds us, Kepler's periodic sentence structure, suspended grammar, and wrenched word order dramatize the slowly building rapture of his long-anticipated discovery of harmonies in nature.

So now since eighteen months ago the dawn, three months ago the proper light of day, and indeed a very few days ago the pure Sun itself of the most marvellous contemplation has shone forth, ... nothing holds me ... (Polanyi, 143)

Kepler experiences the same 'wild surmise' as Keats's 'watcher of the skies' when he sees 'a new planet swim into his ken' or as the stout Cortez when he and his men stare eagle-eyed at the Pacific, 'silent, upon a peak in Darien' ('On First Looking into Chapman's Homer,' ll. 9–14). The bold explorer, however, risks error. Though scientific originality must be 'passionate,' any 'truth-bearing passion is far from infallible.' The exciting prospect of discovery sustains Kepler's intellectual passion through long months of effort. And yet his transforming vision is 'nonsense,' Polanyi says, 'and we would regard it as nonsense today, however accurately it corresponded to the facts; merely and simply because we no longer believe that the fundamental harmonies of the universe are disclosed in such simple geometrical relations' (144).

Without his ardour for scientific discovery, Kepler would never have found and formulated his three laws of planetary motion. But the ardour itself becomes a liability when it prompts him to turn from proposals about geometrical harmonies that happen to be wrong to a hypothesis about a mind residing in the sun or about an astronomer's listening to the different musical scores of the planets, which is not just a false hypothesis but a scientifically incompetent one.

What Thomas S. Kuhn calls 'normal science' is practical and conservative in character: it tries to solve puzzles by applying an accepted scientific paradigm to natural phenomena. More contemplative and radical is the science that explains the phenomena by proposing a new paradigm. If it takes hold, this new paradigm gives rise to what Kuhn calls a 'scientific revolution.' Normal science is a 'tradition-bound activity' which would rather 'suppress fundamental novelties' than subvert 'its basic commitments' (Kuhn, 5). To shatter a tradition, a new theory like Einstein's must reconstruct prior theory and reevaluate established facts. However cataclysmic the revolution may be, it is 'seldom completed,' Kuhn observes, 'by a single man and never overnight' (7).

A successful new paradigm like Aristotle's in his *Physics*, Newton's in his *Principia* and *Opticks*, and Lavoisier's in his *Chemistry* must share two characteristics: it must seem compelling enough to lure adherents away from a competing model; and it must be open-ended enough to leave problems for its new practitioners to solve. Without the constant application of a paradigm, no recognizable research tradition could establish itself or hope to survive. A scientific paradigm takes hold, not because of its contemplative reach or any formal aesthetic appeal it may possess, but because of its practical utility. A paradigm is useful if it promotes the good works of its adherents and justifies their faith in the paradigm's successful application in the future. As Kuhn says, normal science 'is directed to the articulation of those phenomena and theories that the paradigm already supplies ... Mopping-up operations are what engage most scientists throughout their careers' (24). Esoteric problems that fit the paradigm are explored in unparalleled depth. Other problems are ignored until a new paradigm like Einstein's general theory of relativity allows astronomers to study the deformation of space and time in a universe containing massive objects like the sun. As the new paradigm encourages astronomers to make special observations, their wealth of new data allows physicists to test Einstein's paradigm in a setting that is cosmic as well as local.

Normal science is most practical when it uses its paradigm to predict important new discoveries. But since scientists themselves are more concerned with finding new points of contact between a theoretical paradigm and the laws of nature, they usually leave to engineers and technicians the routine work of controlling floods and forecasting storms. Scientists typically ask: What will Newton's new mechanical paradigm tell me about planetary motion, pendulums, inclined planes, tides, or the speed of sound in air? They have less interest in building pendulum clocks or predicting tides. Often, however, the articulation of a theory and its practical application are impossible to separate. As Kuhn concludes, 'the problems of paradigm articulation are simultaneously theoretical and experimental.' Before Coloumb 'could construct his equipment and make measurements with it,' he 'had to employ electrical theory to determine how the equipment was to be built' (Kuhn, 33). His practical application of the theory required him to contemplate it first and refine it afterwards. Often no divorce of practice and contemplation is possible: like Siamese twins, they are joined from birth.

One reason why normal science seems to progress rapidly is that it often concentrates exclusively on puzzles that only a lack of ingenuity prevents its practitioners from solving. Instead of attacking an urgent problem like a cure for cancer, which may have no solution, scientists will concentrate on solvable optical or electrical puzzles that test their ingenuity and skill. 'Though intrinsic value is no criterion for a puzzle, the assured existence of a solution is' (Kuhn, 37). Like a well-played chess game, articulation of an established paradigm will probably make the competitor look good. Since cancer cures may require the invention of a whole new paradigm, a researcher in this area is in danger of looking incompetent when he fails (even though the fame he achieves will be proportionately greater should he happen to succeed).

The differing paradigms of the chemist and physicist produce different answers to the same question: 'Is a single atom of helium a molecule?' Kuhn's chemist says yes, because it behaves like a molecule according to the kinetic theory of gases. The physicist says no, because the helium atom displays no molecular spectrum (Kuhn, 50). In different paradigms the same item acquires different identities. The helium atom resembles Wittgenstein's duck-rabbit figure: now you see a rabbit, now you don't.

Apart from the surprise or shock effect of its solution to a tricky puzzle, normal science never anticipates or takes delight in novelties. When a new fact is found, it is called a discovery: its focus is practical. When a new paradigm is found, we call it an invention: its focus is theoretical. But once again, the distinction between discovery and invention, practice and contemplation, is arbitrary. Kuhn illustrates the interdependence of the two kinds of novelty in the discovery of oxygen by Lavoisier and Priestley. One might say that Priestley discovered oxygen but failed to understand what he found. Lavoisier invented a theory to explain what oxygen is.

Paradoxically, the more precise and far-reaching a paradigm becomes, the more sensitively it can register any persistent deviation from its norms. If a paradigm were less successful, it would also be less attuned to the anomalies that signal the need for change (Kuhn, 65). Unfortunately, a shift in paradigms can be expensive. 'As in manufacture so in science – retooling is an extravagance to be reserved for the occasion that demands it' (76). The costly folly of retooling without just cause is the love of fashion for fashion's sake. It is also the first refuge of ignorance.

Though Kuhn is sometimes criticized for assimilating the paradigms of science to the models and metaphors of art, it seems to me that Kuhn's critics are in greater danger of committing the opposite error of subordinating the poet's imaginative genius to the practical genius of the scientist. These critics come closest to confusing the two when analysing Kuhn's reasons for a scientific revolution. A paradigm cannot be falsified, he argues, 'by direct comparison with nature' (77). Instead, a science on the brink of revolution is starting to compare an old paradigm with the paradigm that may replace it. As Kuhn explains, 'the decision to reject one paradigm is always simultaneously the decision to accept another, and the judgment leading to that decision involves the comparison of both paradigms with nature *and* with each other' (77). In other words, just as Symbolist poets and critics say that literature is always about literature, so Kuhn contends that science is always about science. Its paradigms are always being compared with other paradigms as well as with the natural phenomena they hope to describe more economically and explain more elegantly.

I would argue, however, that Symbolist poetry and art can be only partly assimilated to Kuhn's scientific model. To the degree that science still establishes distinctions and connections with a view to controlling and predicting events, Kuhn continues to recognize that normal science is not autonomous. And to the degree that poetry is a free exercise of an educated imagination, a discipline with its own codebook or grammar of literary symbols, it is not a science. In other words, Kuhn acknowledges an important difference between poetry and science that his critics themselves tend to obscure.

Kuhn believes there are upper and lower limits to scientific research. At the lower limit science confronts what is clearly impossible: the absence of any paradigm at all. No research can be done without paradigms. For 'to reject one paradigm without simultaneously substituting another is to reject science itself' (79). But sometimes a paradigm can be too precisely and comprehensively formulated to allow further refinements. At the upper limit, science paradoxically abolishes itself by becoming too successful. An example is geometric optics, which solves all its problems so flawlessly that it leaves no room for further discoveries. At this point optics ceases to be a subject of scientific research and becomes instead an 'exclusively instrumental' tool for applied use by engineers. Scientific inquiry falls somewhere between these upper and lower limits. Without some sore spots of knowledge, some local resistance to the paradigm, there would be no puzzles to solve. But even points of friction depend on context. What Priestley sees as a solved puzzle in phlogiston theory and what Lorentz sees as a puzzle in the articulation of Clerk-Maxwell's theories, Lavoisier and Einstein both see as counter-instances of the paradigm and hence as trouble spots. The solved puzzle of the 'normal' or conservative scientist gives rise in abnormal or revolutionary science to a wholly new paradigm.

Contrary to what Marjorie Garber says, disciplines like physics and chemistry are not 'constituted ... on the site of their own lack.' The desire of normal science, far from being a 'desire for genius,' is a need to follow given rules or tread familiar paths (Garber, 89). A scientist who examines every tiny anomaly has no time to do significant research. Instead of blaming his tools, he usually renews his effort to remove discrepancies between new observations and familiar theories.

According to Kuhn, only when panic or disillusionment grips a field does science seek a prophet of genius to give a new rule to its order. Just prior to his discovery of relativity, Einstein speaks of the ground being pulled from under him, 'with no firm foundation to be seen anywhere' (Kuhn, 83). And in the months before embracing quantum theory, Wolfgang Pauli admits to being 'terribly confused' and wishing he had 'been a movie comedian' who 'had never heard of physics' (Kuhn, 84). Only such states of crisis can precipitate the kind of 'extraordinary research' in which 'anomaly' is made 'law-like' (85). At such moments scientists like Galileo, Niels Bohr, and Einstein become more contemplative. They may even turn to thought-experiments or 'philosophical analysis as a device for unlocking the riddles of their field' (87).

The invention of a new paradigm is an inscrutable process, usually the achievement of someone 'either very young or very new to the field' (Kuhn, 89). To see that the traditional rules no longer define a playable game, it is an advantage for a scientist not to have become too firmly committed to the rules by prior practice and habit. Though all science is both practical and contemplative, Kuhn argues that normal science is conservative and focused on the discovery of facts: extraordinary science is revolutionary and focused on the invention of theories. The former assumes that science mirrors nature: the latter is a form of science without mirrors. Like his friend Kurt Gödel, a mathematical Platonist, Einstein dreams of uniting gravity and light in a single theory that mirrors no perceivable universe. Perhaps only in a creator-God who is subtle but not malicious can both relativity and quantum mechanics mysteriously coexist. But it may take another physicist and another universe to show how the two theories cohere.

All debate about paradigm choice is necessarily circular. Since an anomaly in one paradigm will become either an axiom or a tautology in another paradigm (a foundation stone that could not possibly be other than it is), logical proof is not strictly speaking at issue. Newton's second law of motion, for example, took centuries of research to discover. But within Newton's own mechanical paradigm, the law that 'an object accelerates in the direction of the net force acting on it' functions more as a logical statement than as an empirical observation. No amount of induction could refute it. Another example can be found in chemistry, where Dalton's paradigm makes the law of constant proportion a tautology, since any reaction in which the ingredients do not enter in fixed proportion is by definition not chemical (Kuhn, 132). 'As in political revolutions, so in paradigm choice – there is no standard higher than the assent of the relevant community' (93). At the practical level of explaining facts, a paradigm has most in common with a scientific experiment. But at the speculative level of inventing theories, the axioms of a scientific paradigm have less in common with an empirical induction than with the articulation of a religious creed or a political manifesto. Einstein's theories of relativity are to physics what the Nicene creed is to Christian theology or what an amendment to the Constitution is to American civil law.

Normally, a physicist or chemist is not in search of novelty as such. Since he wants his experiments to yield the anticipated results, he chooses problems that he thinks his paradigms can clarify and his instruments solve. For this reason, as Kuhn ironically observes, 'an excessive concern with useful problems, regardless of their relation to existing knowledge and technique, can ... easily inhibit scientific development' (95). Pharmaceutical research may be useful, but it may do little to advance our understanding of cellular biology.

Only in a state of scientific crisis will a mysterious phenomenon begin to switch back and forth between different paradigms. If light can sometimes be a wave and sometimes a particle, it is probably neither. The new paradigm of wave mechanics must be invented to create what Kuhn calls a 'self-consistent entity different from both waves and particles' (113). Though Aristotle and Galileo both observed pendulums, and Priestley and Lavoisier both found oxygen, they saw different entities because they viewed the same phenomena through different lenses. A scientific revolution is a kind of gestalt switch. Like Wittgenstein's duck-rabbit figure, it allows science to pass from Aristotle's perception of constrained fall to Galileo's perception of a swinging object. Or it allows Priestley to see dephlogisticated air where Lavoisier sees oxygen. Normal science itself can never correct a paradigm: all it can do is articulate it (Kuhn, 121). A paradigm switch is not an interpretation of new evidence but a flash of intuitive perception that organizes familiar data into new patterns.

A humane discipline that forgot its founders would be convicted of ingratitude and diagnosed with cultural amnesia. But when a scientific paradigm changes, the textbooks in chemistry and physics have to be rewritten. These revisions are carefully designed to disguise the fault lines where cataclysms or earthquakes have shaken apart disciplines like biology and physics. In each rewriting, past theories that no longer contribute to the solution of problems raised by the current paradigm are revised or dropped. Science is the chronicle of winners who distort its history to make their campaigns look progressive and their victories necessary.

Even an individual scientist may make his discoveries appear more cumulative in retrospect than they were at the time they were made. John Dalton, the British chemist, writes as if the combination of proportions were the focus of his chemical atomism from the moment his experiments began. But as Kuhn points out, these combinations occur to Dalton only when his creative work is nearly complete. His own account conceals what is most original in his method: his revolutionary application to chemistry of a set of atomic concepts and procedures 'previously restricted,' as Kuhn says, 'to physics and meteorology' (138). It seems that at moments of crisis and revolution science is suddenly reorganized around new paradigms that require the history of their unanticipated or almost chance discovery to be concealed for strategic reasons.

Kuhn argues that when a highly developed and refined science like geometrical optics solves all its current problems it may abolish itself by turning into technology or a branch of computer engineering. Such a fate is what some commentators call 'the end of science.' At the other extreme, the more speculative or revolutionary a science like physics becomes, the fewer rules there will be to select rationally among alternative paradigms. In choosing a paradigm as one might choose an axiom of faith, science may jettison its experimental method and logic of induction by modelling itself instead on Darwinian biology. As in natural selection, the paradigm possessing those features that allow it to survive by defeating its rivals in open competition will be the favoured model. But since the argument reduces itself to the tautology that what is successful will succeed, the revolutionary scientist also begs the question. He engages in a form of circular reasoning he can never quite conceal.

The celebrated progress of natural science is largely an illusion. By flattening out and distorting its often crisis-ridden history, the textbooks try to make science look more linear and seamless than it is. Since a scientific textbook is always a form of retrospective history, written from the vantage point of the victorious party, it greatly minimizes the cracks and rifts created by the cataclysm. Even when science makes undeniable progress in its solution of carefully selected puzzles, it does so with a rigour that is almost totalitarian. Since the history of science is the chronicle of conquerors and converts, their initiation ceremonies will be as self-celebratory and exclusive as their enforcement of the new regimen is strict. As Kuhn argues, normal science may promote logical brilliance in the manipulation of its paradigm. But its tough discipline seldom develops speculative powers that are of the imaginative depth and intellectual range necessary for the invention of successful new paradigms in the future.

The so-called progress of science is also guaranteed by a natural reluctance to concede that the discarded paradigm explains some puzzles better than the new one. Newton's paradigm could be seen as a loss as well as a gain, since it simply ignores an important question that Aristotle tries to answer in his physics: the reason for the attractive force between particles of matter. Not until Einstein develops his theory of general relativity does physics return to the problem and try to solve it. Even though some scientists initially rejected Newton's dynamics because it failed to explain the attractive forces in nature, Newtonian physics could not openly acknowledge this failure at the time. A seminar on the merits of competing models has little appeal to a science that is still trying to consolidate its revolutionary new paradigm.

Like Newton's dynamics, Lavoisier's chemical theory lost as well as gained in stature when it disallowed questions about the qualities of chemical substances (Kuhn, 148). If science seems to guarantee progress, it is because commitment to a new scientific paradigm is like a religious conversion: a convert is unlikely to concede any defects in the creed he is embracing. Indeed, without commitment, neither a scientist nor a theologian can continue to function. Though Priestley never accepted the theory of oxygen and Lord Kelvin rejected the electromagnetic paradigm, as generations died out the sceptics eventually disappeared until 'the whole profession' once again came to practice 'under a single, but now ... different, paradigm' (Kuhn, 151).

Pure scientists also make progress because they are more sheltered than engineers, medical doctors, and most theologians from urgent social or ethical problems requiring short-term solutions. The substitution of textbooks for original scientific literature also promotes indoctrination. And since a scientific revolution closes 'with a total victory for one of the two opposing camps' (Kuhn, 165), it is highly improbable that a winner will be self-critical or sceptical enough to deny he stands for progress. The scientific enterprise may be fragile and fraught with risk. But as long as it survives, it is bound to seem progressive.

Unlike Karl Popper, Thomas Kuhn does not believe that a scientific theory can be falsified. Even when the results of repeated experiments are anomalous, they cannot be said to disprove anything. Since there is always some imperfection in the fit between data and theories, puzzles that define normal science will continue to arise. But 'if any and every failure to fit were ground for theory rejection, all theories ought to be rejected at all times' (Kuhn, 145). Though a drastic or repeated failure will generate suspicion, the necessary and sufficient cause of testing a paradigm is the presence of some recurrent anomaly or persistent trouble spot – a kind of cancer that infects the whole system. In order to define a falsifying condition, Kuhn believes Popper would have to develop a canon of improbability, including a graduated scale capable of measuring *degrees* of falsification.

All good theories explain the facts, 'but only more or less,' as Kuhn says. 'Though neither Priestley's nor Lavoisier's theory agreed precisely with existing observations, few contemporaries hesitated more than a decade before concluding that Lavoisier's theory provided the better fit of the two' (146). And yet, even though comparison of alternative theories may help determine which paradigm to choose, it is impossible to prove decisively that Lavoisier's paradigm is 'truer.' 'Communication across the revolutionary divide is inevitably partial' (148). And 'like the gestalt switch,' the change to a new paradigm 'must occur all at once (though not necessarily in an instant) or not at all' (149).

Like the proof of a metaphor in poetry or a doctrine in religion, the 'proof' of a scientific paradigm is less a logical demonstration than a testing or probation of it on the proving ground of laboratory experiment and procedure. And yet the scientist's personally chosen paradigm may paradoxically succeed by virtue of its very failure to include the person who chooses it. Whereas Aristotle's physics includes himself, 'it is' (in one philosopher's words) 'this inclusion of the scientist, as he knows himself and as he lives his life, within the scope of his most basic explanations, that modern science [has] renounced' (Sparshott, 1972, 32). Even in exercising personal choice in their discovery and refinement of a new paradigm, Galileo and his successors did something 'extraordinary.' As Francis Sparshott says, they 'formulated a world system in which it was impossible that they themselves could have a place. How could they bring themselves to do something so monstrously silly' (1972, 48)? It is almost as if a freely chosen religion were to exclude the concept of free choice by worshipping a God of absolute caprice and power, like Calvin's God or the voluntarist deity of Augustine.

The switch of paradigms comes as a flash of the right choice to make. It has less in common with the solution of a problem in chess or logic than with Harriet Bulstrode's generous leap of faith in support of her husband Nicholas in George Eliot's novel *Middlemarch*, or with Dorothea's decision to risk further suffering by asking if the humiliation and shame are 'her event only' (549–50, 577). To embrace a new paradigm is to express faith in the future promise of that paradigm: in its ability to explain a far wider range of anomalies than the few problems that the discarded paradigm failed to solve. Kuhn makes the fascinating observation that the anomalies and counter-instances of an obsolete paradigm become the axiomatic first principles of the paradigm that replaces it (78). The scandal of the old paradigm becomes the orthodox cornerstone of its successor, where it can be enshrined as an article of faith incapable of being either proved or disproved. For in contrast to a theorem or a hypothesis, which must always submit to testing, a paradigm that avoids the risk of being proved wrong also forfeits the chance of being proved right. Precisely because it is a paradigm, it is never enshrined like the Decalogue or held to be irreplaceable.

T.H. Huxley acknowledges the importance of both the practical problem-solving functions of what Kuhn calls 'normal science' and the leaps of faith that lead science to embrace a revolutionary new paradigm. He shares Kuhn's belief that only faith in the future promise of a paradigm will lead a scientist to embrace it. Just as Browning's pragmatic Bishop Blougram chooses religious faith over doubt because of its beneficial consequences, so Huxley chooses a materialist paradigm, not because he can prove it is true, but because he finds it more convenient to use. It offers more help to scientists trying to solve difficult puzzles in their discipline.

For Huxley as for Kuhn, science both resembles and differs from religion. As a practical discipline that tests its paradigms, normal science is 'justified,' not by 'faith, but by verification' (Huxley, 1:41). But as a revolutionary discipline that must occasionally replace obsolete paradigms with new ones, science sometimes has to embrace its first principles as a theology embraces the first axioms of its creed. Huxley's appeal to problem solving and utility does not turn science into a technology or a branch of engineering. Science is not an Aladdin's lamp, nor does it exist so that we may 'have telegraphs to Saturn' (1:30). The paradigm shifts in science disclose new and startling mysteries. The Copernican revolution reveals that the earth, far from being the centre of the universe, is a mere eccentric speck. Equally astonishing is the discovery that man, instead of being the crown and roof of things, is a cosmic accident, a biological misfit, whose dreams of harmony tragically unfit him for life and love in a world 'red in tooth and claw,' described by Erasmus Darwin as one vast slaughterhouse.

When accepted as ruling fictions or paradigms, materialist hypotheses can facilitate comparison and usefully extend the frontiers of knowledge. But when mistaken for dogmas, these hypotheses have to be rejected on the same pragmatic grounds as they were once accepted. For when materialism is invested with a misplaced concreteness, it has the pernicious effect of paralysing the energies and destroying the beauties of life (Huxley, 1:165).

Huxley uses such theological concepts as 'justification' and 'faith' to dramatize the truth that the mantle of vatic authority has fallen on the scientist. But theological language carries its own risks. If science's authority lies in the constant testing of its paradigms, then any acceptance of dogmatic authority deprives science of such authority as its 'absolute refusal to acknowledge authority' entitles it to claim. Indeed as Huxley purifies the mind of its materialist paradigms, even as he commends these paradigms as the best way to conduct science, the only authority he accepts is acutely self-critical. If he poses as the enemy of mystification, it is only as a champion of true mystery determined to show that knowledge and bewilderment are compatible. Huxley the scientist is as puzzled as Huxley the religious agnostic. Part of his intellectual equipment is his sense of mystery, his feel for the queerness of nature and the strangeness of her laws.

Huxley's allegiance to both deft calculation and bold theorizing – to the scientist's solution of problems and his invention of new paradigms – leads him to conflate two discrepant pictures. On the one hand, the scientist should function as efficiently as a steam engine or a calculator. On the other hand, he should be full of quixotic fancy and fire, capable of spinning the finest gossamers. The metaphors are as flagrantly contradictory as J.H. Newman's picture of the mind as a spacious reservoir or container (the larger the better), and as a flexibly supple organism. The mind that invents paradigms must be imaginative and subtle. But the patient tester of these paradigms needs different

virtues. He must be as dogged as he is shrewd, and should ideally possess
the logical ingenuity and storage capacity of a computer. Though the
oxymoron of a fiery logic machine capable of spinning the finest
threadwork may possess its own inventive charm, the conceit is patently
absurd. To possess such contradictory qualities is to endure a form of
that acute mental stress which precipitates the mental breakdowns of
Newman and Mill and which makes the routines of ordinary social life
an immense distraction for Einstein.

The scientist's paradigms may come to possess all the disadvantages of
superstition without any of the advantages of religion. A materialist
model like the universe of streaming atoms conjured by Lucretius in *De
Rerum Natura* may have the undesirable effect of inducing more terror
than love in its adherents. But if materialism fills the savage mind with
fear, Huxley believes it is because the hideous idol it denounces is a
phantom of its own making. Though materialism is a spectre to oppress
and appal because it slays the spirit, neither the matter that slays nor the
spirit that is slain has any true existence. The two contending powers are
mere ghosts or fictions. When we demystify the rhetoric, we discover
that matter and spirit are only 'names for the imaginary substrata of
groups of natural phenomena' (1:160). The lament for matter's slaying
of spirit is just as irrational as the lament for Pan. What harm can result
from the death of a fiction?

In defending the invention of bold new paradigms, Huxley opposes
the fallacy of dogmatic authority: the blind reverence that makes the
Ptolemaic paradigm invulnerable to testing or replacement. Like the
axioms of Euclidean geometry, the theological models of the Middle
Ages were postulates of faith that made possible a single world-picture
later dismantled by the Renaissance. But even after the Christian
Humanists recovered new model worlds, they made the fatal error of
assuming that the work of reformation was over. In mistaking the beginning for the end, the Renaissance scholars betrayed the very culture
they recovered. 'The Nemesis of all reformers is finality' (Huxley,
3:149). In his epistles to the philistines, the chief Levite of culture keeps
proclaiming that culture is a criticism of life. Instead of sounding his
trumpets against the walls of science 'as against an educational Jericho'
(3:146), Arnold should have had the intelligence and civility to concede
that 'we cannot know all the best thoughts and sayings of the Greeks
unless we know what they thought about natural phenomena' (3:151).
But just when Huxley seems to have won the argument, convincing us
that scholars have no monopoly on curiosity or criticism, he seriously

damages the case for science by announcing with as much dogmatic
finality as any apostle to the philistines that 'the free employment of reason, in accordance with scientific method, is the sole method of reaching truth' (3:151). Here in all its dangerous arrogance there festers even
in Huxley that cancer of modern learning, the 'research fallacy.'

Truer to Huxley's agnostic temper is his more appealing claim that
science is self-critical because in rejecting dogmatic authority it candidly
admits that all its 'interpretations of natural fact are more or less imperfect and symbolic' (3:150). The merely provisional status of Huxley's
preferred materialist paradigm is confirmed in the twentieth century by
Einstein, who discloses that matter, 'the great bastion of the objectivity
of the world, is an illusion of energy' (Frye, 1982, 14). Indeed the long
footnote in 'The Progress of Science' where Huxley concedes the objections of Boscovich and Clerk Maxwell to a materialist paradigm betrays
Huxley's own anxiety (1:60–1). Even in Huxley's day the dissolution of
matter into fields of energy is preparing the way for Einstein's great shift
of paradigms. In an authoritative scientific work written in 1905, one
commentator, Macneile Dixon, finds 'the sentence, "Ether is the fundamental postulate of physics."' In an equally authoritative work upon
modern physics, dated 1934, the same observer notes that he 'could not
find the word "ether" at all; the "fundamental postulate" did not even
exist in the index' (Dixon, 21). The great enemy of discovery is the pretence of being infallible, the distemper of any high priest of learning
who suspends inquiry whenever some pontiff passes judgment that a
case is closed. 'Roma locuta, causa finita.'

Like scholarship, science nourishes unique qualities of patience,
informed optimism, and humility. As Whitehead observes in *Science and
the Modern World*, the scientific revolution tempers the pride of the scholastic mind by teaching it to bow and submit to nature. Huxley praises a
similar spirit of humility bred in the scientific investigator who has to
subordinate the desires of his mind to resistant data. Jacques Barzun
calls this humility 'a Puritanism of the intellect,' which he praises for its
repression of 'the willful emotions – hope, fear, hate, ambition and self-
deluding joy' (quoted by Bissell, 1968, 201). In addition to humility and
patience, science fosters optimism, a desire to be useful, and a confidence in the mind's ability to control and predict events in the physical
world.

But even in honouring the painstaking testers of paradigms and the
ingenious logicians who solve puzzles with their aid, science cannot
ignore its heroic prophets, rebels, and poets, its inventors of bold new

paradigms. A paradoxical corollary of this need for both puzzle-solving and invention, for practical application and daring surmise, is that any scientist who refuses 'to go beyond fact, rarely get[s] as far as fact' (Huxley, 1:62). Sober science needs its brave guessers, its Keplers and Einsteins. It also needs its long-distance runners, geniuses who combine intuition and stubbornness, lightning flashes of insight with talent for protracted hard work. According to Jurgen Moser, the mathematical genius John Nash 'had no background knowledge' of Reimannian manifolds. Nash's discovery of how to embed these manifolds in Euclidian space 'was totally uncanny. Nobody could understand how somebody like that could do it. He was the only person I ever saw with that kind of power, just brute mental power' (quoted by Nasar, 161). Nash had the daring and simplicity but also the toughness of a 'beautiful mind.' Extraordinarily tenacious, he displayed a prizefighter's capacity 'to continue punching the wall until the stone breaks' (Nasar, 160).

Few warnings are more salutary than Carl Jung's reminder that, although Columbus used 'subjective assumptions, a false hypothesis, and a route abandoned by modern navigation, he nevertheless discovered America' (quoted by Dixon, 253). Even hypotheses utterly at variance with fact, like the geocentric theory of astronomy, can greatly advance scientific knowledge. For error may actively promote scientific discovery by encouraging more accurate observation, more ingenious experimentation, and ever bolder and better speculation. The greater danger lies, not in adventurous surmises or in wild imaginative leaps, but in the formation of hypotheses that are too timid and constrained.

The great defect of Aristotle's theory of science is its separation of the serious study of what is intelligible and therefore teachable in physical nature from the good sense or wisdom of the politician or philosopher. Minimizing the role of bold conjecture in an atomic scientist like Democritus, Aristotle neglects 'the possibility that,' in one commentator's words, 'imagination may have a positive part to play in the life of the mind' (Sparshott, 1994, 210). Though Wordsworth distrusts the analytic mind that 'murders to dissect,' he also knows that imagination is reason in her most exalted mood. Shakespeare's imagination and Newton's intellect are different versions of a single power, which Wordsworth celebrates as 'the breath and finer spirit of all knowledge; ... the impassioned expression which is in the countenance of all Science' (Wordsworth, Preface to second edition of *Lyrical Ballads*, 1950, 688). J.S. Mill shares Wordsworth's suspicion that men of weak imagination are likely to be scientific hacks rather than great trailblazers, since they are impoverished in their fund of fertile speculation. A stunted imagination is a soft-pedal on all spheres of mental activity, not excluding the scientific.

In 'The Study of Poetry' Matthew Arnold calls on humanists to save the world from fact-collecting scientists. The more facts and information science gathers, the more it depletes its stock of original ideas. 'Without poetry,' Arnold predicts, 'our science will appear incomplete; and most of what now passes with us for religion and philosophy will be replaced by poetry' (1961, 306). The harder Arnold tries to confer a specious permanence on facts, the more they dissolve and dissipate in thin air. Like positivism, 'our religion has materialised itself in the fact, in the supposed fact; it has attached its emotion to the fact, and now the fact is failing it' (306). Whereas religion is betrayed by its empirical faith in fact, poetry is saved by its alignment with speculative intelligence or 'ideas.' Indeed, 'for poetry,' he insists, 'the idea is everything.' Arnold's sleight of hand reveals itself in the way he rejects everything in poetry that is not associated with the intellectual and imaginative life as an 'illusion,' even a 'divine illusion.' The aura of illusion that empirical science might be tempted to ascribe to poetry is now ascribed to the positivist's world of sensations and material substances, which turns out to be as insubstantial as the 'supposed fact' that fails religion. As the descriptive meaning of 'fact' begins to change, so does its emotive meaning. But in a desperate effort to recover the original meaning of the word 'fact,' which is the positivist's gold standard, his touchstone of value and truth, Arnold asserts that for poetry 'the idea *is* the fact.' Though Arnold is now using the empiricists' prize word 'fact' in a positive emotive sense, he is also conferring upon it a new descriptive meaning that implies a total repudiation of everything the positivist stands for.

A theory that explains the important place of the scientist in nature is better than a theory that reduces the scientist to an accidental footnote to self-sustaining natural law. A good scientific theory will be mathematically self-consistent. But in order to be consistent it may have to describe an unearthly limbo of nuclear particles – a bloodless world where there are no sounds or colours and no people. A scientific worldview that leaves out the scientist cannot put him back in again. And a third-person world that excludes the scribe who records the beauty, the cruelty, and the mystery is obviously not the world in which any scientist lives. One Nobel Prize–winning physicist, Steven Weinberg, predicts that 'perhaps our best hope for a final [scientific] explanation is to discover a set of final laws of nature and show that this is the only logically consistent rich theory, rich enough for example to allow for the existence' of scientists themselves (50). Any explanation of nature that turns scientists into mere specks of matter deposited by chance on the shoals of time is too spectral and bleak to command widespread assent. 'Though dragged to such conclusions,' as F.H. Bradley says, 'we cannot embrace them. Our principles may be true, but they are not reality. They no more *make* that Whole which commands our devotion than some shredded dissection of human tatters *is* that warm and breathing beauty of flesh which our hearts found delightful' (1883, 533).

4

The Scholar's Knowledge:
The Conversation of the Learned

In a range of splendidly incisive books and essays on higher learning, the English philosopher Michael Oakeshott divides the 'conversation' of liberal education into different 'languages' or discourses. He distinguishes 'the language of the natural sciences,' for example, from 'the language of history, the language of philosophy, or the language of poetic imagination' (1989, 37). A literature stands in the same relation to the language it uses as a learned discipline stands to the models it deploys. To engage in the adventure of university education is to come to recognize and discriminate among what Oakeshott calls the 'languages' or what Mortimer Kadish prefers to call the 'models' of higher learning.

Like Kadish, Oakeshott believes that the grammar of a language cannot be studied in abstraction from the literature in which its genius best displays itself: *The Divine Comedy* in the case of Italian, for example, or *Don Quixote* in Spanish. We must ambush the 'language' of natural science or philosophy by capturing it *en passant*, so to speak, as if the learner were moving a pawn in a chess game in which the genius or presiding spirit of each discipline is the queen the pawn is trying to capture or the king it hopes to checkmate. Despite my homely metaphor, the ambitious purpose of Oakeshott's game is nothing less than the initiation of its players into 'the civilized inheritance of mankind.'

Oakeshott realizes, of course, that few professors are so immodest as to make this claim directly (1989, 49). In teaching the 'languages' of higher learning, professors must *show* rather than *tell* their students how to take possession of the inheritance they were born to. Since the 'vast panorama of futility and anarchy' that T.S. Eliot discerns in contemporary history is likely to present 'a sadly distorted image of a human

being' (Oakeshott, 1989, 48), the professor's first duty may be to impart
a more inclusive sense of history or tradition, one that 'approximates
more closely to the whole of [the learner's] inheritance' (1989, 49).
Such a tradition 'does not deliver to us a clear and unambiguous message,' which would serve only as an ideology or 'crib.' Instead, the larger
tradition often speaks to us 'in riddles; it offers us advice and suggestion, recommendations, aids to reflection, rather than directives'
(1989, 49).

In an overflow of satiric high spirits, animated even for Oakeshott, the
prophet foresees the day when 'a university will have ceased to exist';
when 'those who come to be taught come, not in search of their intellectual fortune,' but only as plunderers and self-promoters (1989, 104). A
university that substitutes the vocational techniques of a business school
or an institute of technology for the 'languages' of science and scholarship betrays its mission. It is as if a parrot or an ape were to mimic the
gestures and accents of a Latin orator.

No author has done more than Oakeshott to characterize the different voices or discourses in the conversation of the learned – historical,
political, and literary. In his ground-breaking essay on 'The Activity of
Being a Historian,' Oakeshott distinguishes between a practical and a
scientific attitude toward the past. To read the past backwards from the
present and to discover in nineteenth-century Romanticism the origins
of modern fascism or communism is to practise a form of 'retrospective
politics.' By contrast, the intellectual historian who refuses to assimilate
Nietzsche or Marx to an interest in contemporary politics assumes what
Oakeshott calls a 'scientific' attitude toward the past. Such a historian
will resist the temptation to study the Romantics' parody of grace and
their appeal to the heart in protest against the dogma of original sin as a
precursor of later communist attempts to build heaven on earth. Seeking the origins of Romanticism in medieval theology as well as in post-
Enlightenment politics, Oakeshott's historian is more likely to see this
cultural movement as a curious secular version of the Pelagian heresy,
which rejects the doctrine of the Fall. The intellectual historian loves
discarded ideas 'as a mistress of whom he never tires and whom he
never expects to talk sense' (1991, 182).

According to Oakeshott the historian's task is 'to translate action and
event from their practical idiom into an historical idiom' (1991, 180).
'Concerned with the past for its own sake and working to a chosen
scale,' the historian 'elicits a coherence in a group of contingencies'
(1991, 183). Since 'each genuine piece of historical writing has a scale'

or decorum 'of its own and is to be recognized as an independent example of historical thinking,' it is better for historical scholarship to be written in the carefully proportioned literary idiom of a poem or a novel than in the scientific idiom of a laboratory report or in the practical idiom of a newspaper editorial, which is too often a retrospective exercise in partisan politics. To be sure, history is not poetry and scholarship is not prose fiction. Oakeshott is merely saying that in historical writing 'concessions to the idiom of legend' are 'perhaps less damaging than other divergencies' (1991, 183).

In *The Voice of Liberal Learning* Oakeshott praises history for its power to give cultures possession of their heritage. Along with literature and the other arts, history is a 'singing school' for the soul, a study in what Yeats calls 'monuments of its own magnificence' ('Sailing to Byzantium,' ll. 13–14). If greatness is simply a secular name for divinity, then history like theology must create a theatre in which citizens can observe a stirring national drama. By allowing its scholars to see their world more inclusively, history allows them 'to understand and become possessor[s] of' the legacy they are 'heir to' (1989, 48).

To become an appropriate subject of higher learning, politics must learn to disengage itself from politics proper and accept the challenge of trying to master the explanatory discourse of a whole archive of historical and philosophical documents. The language of politics is the most difficult to handle, Oakeshott thinks, because 'the idiom of the material to be studied is ever ready to impose itself upon the manner in which it is studied' (1991, 218). Paradoxically, a 'School of "Politics" ... should never use the language of politics; [it] should use only the explanatory "languages" of academic study' (1991, 216). Oakeshott admits that 'the words which compose our vocabulary of politics' will probably 'be uttered.' But the only justification for referring to democracy or liberalism will be 'to inquire into their use and meaning, in order to take them to pieces and write them out in the long-hand of historical or philosophical explanation' (1991, 216).

According to Oakeshott, political treatises have more in common with recipes in a cookbook than with the axioms of Euclid's geometry. Political theory is less a *terminus ab quo*, a launching pad for future social experiment, than a *terminus ad quem*, a summary of past political practice and experience. An ideology like the Rights of Man, for example, which stirred the minds of revolutionaries in eighteenth-century America and France, 'no more existed in advance of' the day-to-day exercise of civil rights, spanning many centuries in England, 'than a cookery book exists in advance of knowing how to cook.' And 'so far from being a preface,' Locke's *Second Treatise of Civil Government,* venerated in America and France as a revolutionary creed about to inaugurate a new Utopia, 'has all the marks of a postscript, and its power to guide derived from its roots in actual political experience' (1991, 144).

In Oakeshott's view, however, the most contemplative mode of discourse in higher learning is not history or politics but poetry and literary criticism. Instead of aspiring to the 'voice of argumentative discourse,' which is the voice of 'science,' Oakeshott believes that higher learning should be a conversation in which 'no proposition' is 'to be proved' and 'no conclusion sought' (1991, 489). Though the idol of Oakeshott's cave may be too close for modern taste to the refined talk of an Oxford common room, there is much to be said for his ideal of an 'unrehearsed intellectual adventure' in which different voices can be heard participating (1991, 490). To relieve the boredom of a conversation in which the familiar voices of practical activity and science dominate, Oakeshott recommends a return to poetry and criticism, discourses which restore the serious play of a reflective mind.

A chemist who analyses a sample of water for traces of *E. coli* bacteria has a scientific interest in the water. A medical doctor who prescribes water as a cure for dehydration has a practical interest in the compound. Furnishing a contrast to these scientific and practical uses of H_2O is Keats's image of 'the moving waters at their priestlike task / Of pure ablution round earth's human shores' ('Bright Star,' ll. 5–6). As a 'thing of beauty' and 'a joy forever,' such a disinterested, contemplative use of language has an odd affinity with friendship and love. Unlike practical concern with flood control or an agricultural interest in irrigation, a friend is not someone who has useful qualities. Instead, 'he is somebody who evokes interest, delight, unreasoning loyalty, and who (almost) engages contemplative imagination' (Oakeshott, 1991, 537).

Like Philip Sidney's poet, who never lies because he never affirms anything, Oakeshott's poet betrays his vocation only when he aspires to logical proof or demonstration. 'To give an accurate description of what has never occurred,' Oscar Wilde suggests, is 'the proper occupation of the historian' (1968, 206). Though it is not 'the proper occupation' of Oakeshott's historian, it is certainly the 'inalienable privilege' of his poet. For poetic images are like voices in a conversation, which 'neither confirm nor refute one another' (Oakeshott, 1991, 540). The moment poetry 'imitates the voice of practice its utterance is counterfeit' (1991, 540). Though there is always a man from Porlock to wake even the most reclusive scholar from his reveries, poetry is 'a sort of truancy, ... a wild flower planted among our wheat' that rescues 'from the flow of curiosity and contrivance' (1991, 541) a few fugitive moments of re-creation and insight. In practical life poetry also cultivates in its admirers friendship, love, and an affection for that 'conversation of mankind' which Oake-shott values as one of learning's keenest pleasures.

In Matthew Arnold's view, a polemical criticism of poetry is absurd and contradictory. For true criticism is by definition disinterested, a 'free play of the mind on all subjects which it touches' (1961, 246). To be disinterested or pure is not to be indifferent but to remain independent of the partisan politics of literary journals and reviews. 'Practical considerations cling to [the mind] and stifle' any free exercise of thought. When such partisanship prevails, criticism of poetry or any other art can never be practised in a truly liberal service of the self. For the party spirit is immune to logic and criticism. At its most self-damaging it refutes itself in an outrageous miscarriage of thinking: 'If one of us speaks well, applaud him; if one of us speaks ill, applaud him too; we are all in the same movement, we are all liberals, we are all in pursuit of truth' (Arnold, 1961, 251). If we substitute for Arnold's partisan use of 'liberal' such recently fashionable labels as 'feminist,' 'postcolonial' or 'gay,' we can appreciate the timeliness of Arnold's attack. A partisan appeal in criticism masks in vain multiple failures to discriminate and think. Too often the blundering and illogical pursuit of truth is mistaken for its attainment. And since incompetence and ignorance are rewarded as generously as competence and knowledge, there is neither the capacity nor the incentive to evaluate. If a then b; if not a, then b, too. Since b is the consequence of both a and the opposite of a, it can be a genuine effect of neither. Truth disappears in the very act of describing how we pursue it. And to make the outcome as dark and calamitous as possible, Arnold quotes Joubert's aphorism: 'Ignorance, which in matters of morals extenuates the crime, is itself, in intellectual matters, a crime of the first order' (1961, 252).

Like poetry, literature in general, I believe, is just as cognitive as science. But it transmits a different form of knowledge by replacing the sequential arguments of ordinary reasoning with such presentational forms as parables and aphorisms, whose force must be grasped all at once or not at all. The truths of great art are truths of embodiment, which it offers as an alternative to truths of logical demonstration and rebuttal. What can be shown but not said in art are not just academic questions but shocks of encounter that demand some response of risk or commitment, since they often raise life-and-death issues we must deal with.

Some theorists would argue that a contemplative literary criticism is a contradiction in terms. Scholars who contemplate a poem are apprehending it as a whole. But when they analyse it, they divide it into parts: in Wordsworth's phrase, 'they murder to dissect.' Even prophetic or visionary critics like Geoffrey Hartman and Northrop Frye, who are more contemplative than most contemporaries, have been accused of contradiction. Ideally, Frye combines critical judgment with prophetic wit. But how can he pretend to be a discriminating judge of difference at the same time he proclaims the unity of the Word?

According to Denis Donoghue, Frye's '*Anatomy of Criticism* is the supreme Book of Distinction, in which every type of literature, every genre, is shown defending its turf' (1992, 25–6). In Frye's oracular pronouncements about the unity of a Logos, 'we hear much of community.' But even as Frye poses as a prophet and a wit, what he shows us as a discerning judge of cultural difference is 'a proliferation of tribes, each of them maintaining itself by reciting the stories of its gods' (1992, 26). Donoghue thinks it is contradictory for Frye to 'invoke, at one moment, a vision of human unity, founded upon story, fiction, and metaphor; and at the next moment set up partitions between one community and another' (1992, 25). Though Frye is remembered for his witty digressions and playful celebration of difference, he is also a monist of supple obduracy – a Joshua whose trumpet blasts cause many walls to tumble down. A similar fate awaits T.S. Eliot's mentor, F.H. Bradley, who sings a serene Vedantic hymn to the Absolute in the second part of his metaphysical essay *Appearance and Reality*, but who is best known today for his brilliant deconstruction of the world of appearances, which is the grave that swallows up his own avowed Idealism.[1]

Some critics of efforts to integrate big cultural concepts in a canon of Great Books reject the study of Aeschylus and Shakespeare because they think these authors celebrate Western hegemony. But as David Denby says in their defence, the Western classics 'are less a conquering army than a kingdom of untamable beasts, at war with one another and with readers' (460). We read the classics, not because we are cultural imperialists who want to integrate the world coercively, but because such works provide 'the greatest range of pleasure and soulfulness and reasoning power that any of us is capable of' (463). Far from being the playground or preserve of a privileged few, the great classics from Homer to Shakespeare help us make sense of those acts of loyalty and love and those breaches in trust that define who we are. Biblical literature tries to rectify the pain and mystery of our lives by using God's covenant with Noah or the crucifixion as a marker of the place between justice and mercy, death and life, where most people find themselves. Without denying the dignity of difference, such literature is acknowledging that storytelling or myth is the bond of our shared humanity.

Naturally, ideological critics, political activists, and advocates of Third-World literature can be just as two-dimensional or arrogant as the Westerners they attack. As Jeffrey Hart observes, their 'villain always turns out to be variously white, male, Western, racist, imperialist, sexist or homophobic – or, with luck, all of them together. The result is ... not literary experience but an endless repetition of slogans and clichés. Careful reading of important works resists and undercuts these muralistic cartoons' (246). Instead of enforcing the prejudices of our own class or time, the great books I taught for forty years develop 'self-critical habits of mind' that offer 'less a triumphant code of mastery' than what David Denby calls 'a tormented ideal of obligation and self-knowledge' (145). In *A Passage to India*, for example, E.M. Forster allows his characters to cross boundaries of race, gender, and age. But by putting each of these experiments of living under strain and pressure, Forster also asks readers to consider whether such utopian experiments are out of touch with the real world, or whether the world itself ought to live up to such exalting imaginations of it.

In Mark Twain's greatest novel, *Huckleberry Finn*, social forces both on and off the raft conspire against the youths' brave experiments in male bonding and friendship. When Huck helps his friend Jim escape, even if it means 'going to hell,' he is struggling to express revolutionary moral insights in the language of a southern white supremacy that his boldest intuitions challenge and subvert. The great changes in gender relations and in treatments of love, friendship, and male and female bonding take place first, not in political manifestos like Mary Wollstonecraft's *Vindication of the Rights of Woman*, but in poems, short stories, and novels that challenge the authority of cultural norms.

A signal achievement of programs in Women's Studies is their appreciation of the power of genre to construct and shape a sense of gender and identity. Experiments in genre are often 'experiments of living' in which such nineteenth-century women novelists as Charlotte and Emily Brontë oppose to male authority a feminist counter-tradition in English fiction. At the same time women poets like Elizabeth Barrett Browning, Felicia Hemans, and L.E. Landon reinvent the dramatic monologue as a way of opposing and controlling the male objectification of women. Great literature often bequeaths a fierce legacy of self-division and conflict. It may leave young students badly hurt precisely because it challenges rather than confirms many of the preconceptions they live by.

According to William James, any subject can be studied contemplatively and humanely if it is studied historically.[2] The *OED* is to the history of the English language what Darwin's *Origins of Species* is to the history of life: for static synchronic study it substitutes a dynamic story of accidental variation and change. Charles Lyell's new geology of universal flux is a solvent. But even as Darwin uses it to dissolve traditional taxonomies, he uses the concept of a species to stabilize the flux and give the story of plant and animal development an intelligible history or plot.

Conversely, the contemplative scholar as literary critic acquires something of the scientist's talent for discovery when he discerns unsuspected similarities among disparate works of literature. Sometimes Tennyson's most resonant allusions are to other allusions, like Bedivere's echo of Milton's echo of Virgil. Harking back to the narrator's self-correction in 'Oenone' – 'and round her neck / Floated her hair or seemed to float in rest' (ll. 17–18), Tennyson's claim that Bedivere 'saw ... /Or thought he saw' King Arthur's boat ('The Passing of Arthur,' ll. 463–5) summons up Aeneas's seeing the dim form of Dido through the shadows of the Virgilian underworld, 'as early in the month one sees, or thinks he has seen the moon rise amid clouds' ('aut videt aut vidisse putat,' *Aeneid*, 6.454). But even as Tennyson's little insect-dance of hesitation recalls the moment in Virgil when Aeneas glimpses his cherished but ill-used Dido, he also evokes and reinvents the most important earlier use of this Virgilian allusion in English poetry. For behind Virgil's simile we hear the diminishing epic simile Milton uses in *Paradise Lost* to reduce the conclave of demons to elves seen at midnight.

> or Faerie Elves,
> Whose midnight Revels, by a Forrest side
> Or Fountain some belated Peasant sees,
> Or dreams he sees ...
> *Paradise Lost*, 1:781–4

The literary critic also displays a scientist's talent for testing and verification when, instead of travelling back through the history of a genre, he shows how a resourceful elegy like *In Memoriam* crosses genres by combining the lapidary concision of an epitaph with the trajectory of a confession spanning long intervals of time. Discovery is a form of wit-criticism (disparate things are alike); while practical testing is judgmental (each work of art is irreplaceable and unique). Though Frye himself is a great critic of genres, all generic criticism founders on Frye's aphorism that 'comparisons are relevant only to inorganic quantities and that all qualitative units such as human souls [and he might have added 'works of art'] are incomparable' (2000, 27).

Oscar Wilde wittily dismisses learned conversation as 'either the affectation of the ignorant or the profession of the mentally unemployed' (1968, 206). But by conversation Michael Oakeshott means something less idle or pretentious. He is referring to a civilization's cultural inheritance and its gift to posterity. Not to be mistaken for a collection of tapes in a museum or for a stack of manuscripts in an archive, the literary, historical, and philosophical conversations of the learned compose what Oakeshott calls a capital of discourses – of manners and methods of thinking – to be used and expanded from one generation to the next. Developing the economic metaphor of a culture's stock versus its capital, Oakeshott notes that when capital is used, 'it earns an interest.' The part of this interest income 'which is consumed in a current manner of living' is the income a culture spends on its professional and vocational training (1991, 187). The portion of its interest that a culture reinvests in speculative scholarship and theoretical science is the increase in capital that it hopes to pass on to posterity.

Oakeshott argues that the more enlightened a culture, the greater the fraction of its intellectual endowment it will invest in the capital formation of disinterested scholarship and science. The more material it is, the larger the portion it will use up in immediate consumption and technology. Perhaps the Middle Ages was more enlightened in this respect than most contemporary cultures. James Cameron is astonished by the large percentage of its modest surplus medieval society invested in higher learning. 'The amount of the wealth of [this] society given to the universities must have been immense,' he marvels; 'one hardly knows how the economies of Ireland and Sicily and Bohemia and Scotland could have spared the many young men who went off to Oxford or Paris or Bologna or Cologne, often enchanted at first by the music of dialectic, though no doubt a great many stayed to study law, canon and civil, the great path to advancement in such a society' (30). Regrettably, a culture that worships technology and commerce distrusts the more speculative knowledge that comes from a mastery of games like chess or from contemplative pursuits like mathematics and philosophy. In Oakeshott's words, the models of such disciplines 'can neither be taught nor learned, but only imparted and acquired' (1991, 15). There is a danger that in failing to reinvest a portion of its wealth in scientific discovery and disinterested scholarship, a culture that spends most of its intellectual resources on applied knowledge will rob posterity of its inheritance. Future generations will have less theoretical knowledge and less wisdom to apply.

If the university is to care for 'the whole intellectual capital which composes a civilization' (Oakeshott, 1991, 194), then most of its earned interest should be reinvested in developing a wide array of 'languages' or models in carefully selected disciplines. A university's most valuable contribution is not the mere knowledge it acquires or the information it imparts. More important is the training it provides in the disciplined use and bold reinvention of models. And yet society's own interest in these models is likely to be short-term and pragmatic. A medical doctor may be concerned only with such recent results of biochemical research as offer a cure for genetic heart diseases. Instead of learning to speak the language of the medical scientist, the clinician will merely acquire skill in reading and using the information contained in the genetic code. For the computer programmer, as for the cardiologist who is not engaged in active research, 'the world began the day before yesterday.' As Oakeshott says, 'what is learnt is not a "living" language with a view to being able to speak it and say new things in it, but a "dead" language; and it is learnt merely for the purpose of reading a "literature" or a "text" in order to acquire the information it contains' (Oakeshott, 1991, 192–3).

The loss of diversity in the conduct of scholarship or science is like the extinction of a species or the loss of a language. Though no generation ever transmits to posterity everything it knows, a sudden disappearance of diversity often signals intellectual decline. Specialization is often the price of progress. But in scholarship, as in agriculture, it is dangerous to rely on one resource alone. Ireland's famine in the mid-nineteenth century was a result of planting only one species of potato. When sugar beets succumb to parasites, it is always useful to have sugar cane as a substitute. A short decade ago Victorian literary scholarship in my own university was represented by a rather imposing array of biographers, textual editors, intellectual historians, and critics. For two generations the Victorian scholars refused to produce clones of any single group within their membership. Recently these scholars have been turning into replicas of one small band of cultural historians. Though approaches to literary studies are dividing faster than amoebas, the latest fashion is not always the most inclusive or enduring. Even the novelties of the New Historicism are only the resuscitated fashions of the generation before my own. Ironically, the recent disappearance in some British universities of entire departments of classics and philosophy has started to erase the memory of even Socrates himself. It is obliterating our knowledge of the very institutions to which universities trace their origin, the ancient academies of Plato and Aristotle.

To support diversity in the 'conversation of mankind' a department should encourage mutual respect among its members. It can foster conversation by providing more opportunity for faculty and students to share ideas. Despite what many administrators seem to think, original scholarship is not amenable to mass production methods. My chairman expressed alarm last year when only four members of the department submitted recently published books for display at the Book Fair. But when a shrinking faculty is forced to teach larger classes, direct more theses, sit on more committees, and read endless scrolls of e-mail, little time remains for reflective scholarship. Just as scientists require well-equipped laboratories and an efficient corps of technical assistants, so scholars need a generous supply of research leaves, faculty lounges, reading rooms, and conference facilities. Informal gatherings that allow colleagues to discuss their work in progress help older scholars keep abreast of new developments in feminist criticism and semiotics. It also helps the 'theorists' purge their discourse of unnecessary jargon, which is a conspiracy not only against the laity but also against other literary scholars. Such groups do wonders to ensure that no prevailing method or vocabulary is allowed to set up barriers between scholars working on related projects.

The discourses that enter into Oakeshott's 'conversation of mankind' should be rich and varied: like the description of a lecture once given by Emerson, the different conversations may exhibit no discernible connection except in God. Indeed any monolithic notion of what a department is expected to do is likely to breed dissension and paranoia. Ideally, a few inspired souls will be moved to build bridges between old and new ways of doing things, since it seems pernicious to declare one 'conversation' central and others peripheral. What seems at the margin today may move to centre-stage tomorrow. And any declared centre is usually a still centre. Since undergraduate students tend to gravitate to fashion, however, knowing little else, a department that hires only feminists, Marxists, or cultural historians will soon be obsolete. There is no substitute for scholars who are thoroughly grounded in the traditional methods of their discipline, yet flexible enough to assimilate new developments in the field. Alexander Pope's warning is always timely: 'Be not the first by whom the new is tried / Nor yet the last to lay the old aside' (*An Essay on Criticism*, 2:335–6).

Taking part in Oakeshott's 'conversation of mankind' is more than playing a role in an academic charade. Too often we regard interdisciplinary studies the way Nietzsche says the nineteenth-century regarded history, 'as a storage room for costumes' (150). Though education should encourage scholars to conduct 'experiments of living' and even cultivate the negative capability of a gifted playwright or actor, it should also remind each born philosopher, chemist, or mathematician that there is 'something unteachable' in him, what Nietzsche calls 'some granite of spiritual *fatum*' (162), which he alters at his peril. Misguided counsellors have turned too many gifted writers into second-rate lawyers, and have made mediocre scholars and scientists out of too many talented entrepreneurs who should have been in business. The history of vocational success is often the history of sturdy individualists too obdurate to be sculpted into Pygmalion's statue. They owe such genius as they possess to 'the great stupidity' they are, to 'what is *unteachable*' in them 'very deep down' (162). We must learn when to reinvent ourselves by trying to be original and when, by taking part in too many conversations,[3] we become a mere parody of ourselves – one of God's jokers or buffoons.

At the risk of forfeiting a genuine identity of his own, a talented actor will acquire a repertoire of dramatic parts, just as a gifted writer of dramatic monologues will imitate a ventriloquist or a whole troupe of different speakers. A university that gives its students practice in a wide array of discourses may be little better than an acting school that produces more chameleons than scholars. In my view the major risk of role-playing is that it may paradoxically minimize the risks of self-discovery by allowing an aspiring scholar or scientist to substitute imitation for invention. The dexterity or flair of a scholar who masters several academic models should never be confused with the mere versatility of an actor or the *savoir faire* of a dilettante. The experiments in living of a social self are experiments in negative capability: the student who writes in many styles or who tries his hand at many disciplines still lacks an authentic commitment. To acquire flair in writing a literary essay is not synonymous with mastering the great organizing models of literary scholarship and criticism. And to conduct with dexterity a single experiment in physics is not synonymous with thinking and behaving like a physicist. It is easy to mime the mannerisms and routines of a discipline, but much harder to master, apply, and some day even transform its great models or paradigms.

Science fosters humility, patience, optimism, a desire to be useful, and a confidence in the mind's capacity to control and predict events. A scholar, by contrast, values reflective intelligence and wisdom. He tends to rank unsystematic brilliance above patience and humility. Replacing utility with self-growth, the scholar finds more to admire in the disinterested curiosity of Arnold's cultured humanist than in the shrewdness of a business executive or a crafty politician. The scholar is less impressed by utility and rapid economic expansion than by the 'effort of a cultural tradition' to maintain 'continuity' and to 'check and control' what F.R. Leavis calls 'the blind drive onward of material and mechanical development' with its disastrous 'human consequences' (1943, 16–17).

Leavis believes the challenge of higher learning is to 'produce specialists who are in touch with a humane centre, and to produce a centre for them to be in touch with' (1943, 28). Such is the function of contemplative scholarship in an age of science. It educates the scholar 'in that school of unspecialized intelligence which is created in informal intercourse – intercourse that brings together intellectual appetites from specialisms of all kinds, and from various academic levels' (1943, 28).

But how is Leavis to assimilate or explain the difficult logic of an unspecialized specialist? If learning is too specialized it is not humane, and if it is too unspecialized it is not knowledge. In literary scholarship and criticism, models must be mastered. But even as intellectual rigour provides the necessary discipline, cultivation of imagination and sensibility has to supply a humanizing centre. The qualities to be sought in an unspecialized specialist are a combination of 'intelligence and sensibility.' The educated humanist, like the gifted literary scholar, must cultivate 'a sensitiveness and precision of response and a delicate integrity of intelligence – intelligence that integrates as well as analyses and [that] must have pertinacity and staying power as well as delicacy' (Leavis, 1943, 34). Though such qualities seem at times to contradict each other, like Newman's portrait of the Oxford gentleman, Leavis argues that they alone are capable of integrating all branches of scholarship by providing 'the active informing spirit of the whole' (1943, 34).

Unless philology and language training are combined with an 'appreciative habituation to the subtleties of language' in great works of literature (1943, 38), Leavis believes a literary education 'will defeat its

avowed and traditional ends – those of liberal education' (1943, 39).
But like life itself, learning to write and be fully human 'is a continuous
process for which credit is an illusion and competence an inadequate
term' (1943, 55). As Hazard Adams says, 'no one ever believes he writes
well enough unless he has what, in my opinion, is an inadequate measure of desire' (55). Moreover, liberal scholarship, wherever it exists,
can never be viewed as a single spoke of the academic wheel. As Adams
insists, 'the proper arrangement of the disciplines in higher education
is vertical, not horizontal' (51).[4] Since reasoning is one of the many
operations we perform with language, writing that is lucid and supple is
not just an elegant acquirement but the means of conscious life. It is the
foundation or cornerstone of everything we build.

The purity of poetry and its freedom from corruption justify the central place of literary criticism in higher learning. Poetry is pure because
it resists manipulation.[5] It cannot be made to mean something other
than itself. Though poetry is conspicuously *not* useful, it has an important educative function. In training the mind to recognize patterns and
to take a disinterested, inclusive view, literary criticism cultivates many
of the same contemplative values as inspired teaching. When describing
the progress of Hallie's lecture on Keats in her novel *The Small Room*,
May Sarton compares the conduct of a literary discussion to the development of a fugue.

> And slowly, what had been a painful, stumbling series of unrelated ques-
> tions and answers became something like a fugue. Hallie was gently impos-
> ing a line, bringing them back to certain themes played over and over –
> thought, language, character, the making of a poet. And as she led the class
> back to these major chords, again and again, weaving in and out, what had
> in the first few moments been a professor 'drawing' out a student, had
> become now a true dialogue. (113)

In Sarton's memorable praise of the mind's delight in the display and
enjoyment of form for its own sake, the withdrawal (not only of poetry)
but of teaching itself into a segregated space is reinforced by a tendency
to think of both art and religion as a *via contemplativa*, a temporary
refuge from the confusion of everyday life. Like a liberal education, a
devotion to teaching for its own sake yields such cognitive pleasures as
identifying fugue-like patterns, deciphering symbols, and appraising the
shock effect of encounters that are both surprising and well prepared.
Sarton also discloses an unexpected similarity between the informed

sympathy of an intelligent class discussion and the values represented in a great poem.

The development of a 'stumbling series of unrelated questions' into 'something like a fugue' is most concentrated in a work like *In Memoriam*, where both the substance and spirit of section 36, for example, follow from what precedes, and where the need to hold each element in more than one context helps the mind enlarge its vision and assimilate its dread.

> Though truths in manhood darkly join,
>   Deep-seated in our mystic frame,
>   We yield all blessing to the name
> Of Him that made them current coin;
>
> ...
>
> And so the Word had breath, and wrought
>   With human hands the creed of creeds
>   In loveliness of perfect deeds,
> More strong than all poetic thought;
>
> Which he may read that binds the sheaf,
>   Or builds the house, or digs the grave,
>   And those wild eyes that watch the wave
> In roarings round the coral reef.

There are usually openings at the centre of any active section of *In Memoriam*, moments when sharp breaks inside a wavering body of sounds – 'that binds the sheaf / Or builds the house, or digs the grave' – set up a disturbing tremor just before the mind is dazzled by a daemonic vision. For a split second we are inside the mind of a wild poet, on the verge of madness, who passes unpredictably into a world of extravagant surmise. At first the links are reinforced by the economic metaphor of line 4 – 'Of Him that made them current coin' – which answers the earlier question: 'what *profits* it to put / An idle case?' (35.17–18). The connections are also strengthened by line 12 – 'More strong than all poetic thought' – which casually picks up the contrast between aesthetics and theology that has been central to the conceit of the 'wild Poet,' God, working 'Without a conscience or an aim' in section 34 (ll. 7–8). Do we hear Schopenhauer or Wordsworth? A brooding Hardy or a benign

John Keble? Despite the predictable tetrameter quatrains, the tone is remarkably active and varying. It keeps inviting tiny adjustments of feeling. The roaring waves 'round the coral reef' in section 36, l. 16, are just as incessant as 'The moanings of the homeless sea' in section 35, l. 9. The only difference is that the watching eyes can now perceive a design and final cause in nature. The beauty of the reef is now muted and incidental; Tennyson subordinates it to the 'loveliness of perfect deeds,' a theological mystery 'More strong than all poetic thought' (section 36, ll. 11–12). He also ascribes the wildness to a creature beholding God's creation, whereas in section 34 he attributed it directly to God himself. Tennyson wants to go so far with his metaphors and no further. As Frost says, 'all metaphor breaks down somewhere. That is the beauty of it. It is touch and go with the metaphor, and until you have lived with it long enough you don't know when it is going (1968, 41).' The reader never knows how much meaning he can get out of these fugue-like developments and when they will cease to yield. In the midst of death, they are like life itself.

This is true even in section 10, where Tennyson's impulse to scrutinize and be critical coexists with a readiness to begin a lyric with a trite account of foreign travellers and a sailor returning to domestic comforts.

> Thou bring'st the sailor to his wife,
>   And travelled men from foreign lands;
>   And letters unto trembling hands;
> And, thy dark freight, a vanished life.
>
>   10.5–8

A letter received with 'trembling hands' is probably no ordinary letter. And the 'dark freight' of a 'vanished life' has power to alarm and unnerve the reader, perhaps because it is a conceit that is only half intelligible. As a fictional character explains in A.S. Byatt's novella 'The Conjugial Angel,' 'the *weight* of the freight, so to speak, is the weight of absence, of what is vanished, a lost life. It is not what remains that is heavy, but what is not there' (1992, 235). For free rational minds there is no higher sanction by which an education of this kind can be justified. Unfortunately, it is easier to be trained in a theory or schooled in a cause than educated by poetry.

Like Virgil, Tennyson combines fine nuances of feeling with an energy that is implacable and daemonic. Harold Bloom thinks that Virgil's deep male distrust of female power expresses itself in his portrait of the goddess Juno, a monster and a demon. A comparable fear pervades the elegiac eros Tennyson associates with Vivien and the beautifully destructive Maud. In his monologue 'Lucretius' Tennyson represents Venus as a blind swerve of atoms ruining through an 'illimitable inane' in an unearthly dance that even the Furies would think twice before choreographing. Most appalling is *In Memoriam*'s consistent personification of random force as a demonic harpy or female demon, like the Sibyl 'with raving mouth' in Heraclitus, who utters 'solemn, unadorned, unlovely words' that sound and echo for 'a thousand years.'

> From scarped cliff and quarried stone
> She cries, 'A thousand types are gone;
> I care for nothing, all shall go ...'
> *In Memoriam*, 56.2–4

As Tennyson becomes a scribe of the 'terrible Muses,' the new astronomy and geology, a Heraclitean science of universal flux leaves 'man,' God's noblest work, stranded for three successive quatrains without a verb. Tennyson's dark daemon, the fierce harpy, shrieks from the cliff and initiates a frantic tripling of relative pronouns: 'Who rolled ... / Who built... / Who trusted' [ll. 11–13]; 'Who loved, who suffered ... / Who battled' [ll. 17–18]). When this 'roof and crown of things' ('The Lotos-Eaters,' l. 24) starts to stumble and weaken, there is a cry for Hallam (the subject of the elegy) to dissolve the mockery. The verse combines a fine attention to varieties of human purpose and achievement with a shattering, annihilating indifference to them all. Repetitive contrasts and accumulative doubts mount in a crescendo of fear that is typical of the manic phase of depression and melancholia. A moment later they subside into fragments without verbs – 'No more? A monster then, a dream, / A discord' (ll. 21–2), which all but suspend the neural flow. What is most Virgilian about Tennyson is the way the inexorable energies hold sway, annihilating everything of value, even while the poet himself, as Harold Bloom says of Virgil, 'is exquisitely susceptible to every anguish he portrays' (2002, 78).

J.H. Newman's Oxford gentleman, a cultivated aesthete and humanist, is no longer the university's ideal product. And in the modern stereotype of the pleasantly dithering, eccentric professor, it is easier to find a caricature of the scholar than his authentic avatar. More practical and attractive, I think, is May Sarton's Hallie, the inspired lecturer on Keats, or the unspecialized specialist whom F.R. Leavis praises as his model of the literary scholar. The best lecturer on poetry does not merely carry her discourse to *predictable* conclusions. Like Sarton's Hallie, she realizes that the outcome of any complex inquiry is unpredictable. Instead of talking in short, non-sequitur fragments, the gifted lecturer takes time to develop her ideas by moving in circles. Rising to the challenge of giving shape to a 'rather loose excitement' by improvising an artful Socratic 'fugue' in which 'master and pupil' become 'partners' in a dialogue of one (Sarton, 116), Hallie brings curiosity to her inquiries, staying power to her interests, and a civilizing intelligence to her conversation with students.

Plato speaks highly of philosopher kings. But academics like Woodrow Wilson seldom make good politicians. And able statesmen like Franklin Roosevelt are seldom scholars. As Bertrand Russell says, 'the trouble with the world is that the stupid are cocksure and the intelligent are full of doubt' (quoted by Charlton, 43). According to Goethe, the man who acts decisively is always without scruples. The counter-truth is that a reflective intelligence is a product of scholarly introspection and self-doubt. Professors are often too Hamlet-like in their scruples to act effectively. Even as a subversive social critic, the scholar tends to be a comic rationalist like George Bernard Shaw rather than an activist like Lenin. Instead of mixing ideas into gunpowder, he allows his revolutionary paradoxes to blow up harmlessly in word play or jokes.

The scientist's scholarship of discovery seldom contains degrees of depth or profundity. For such depth we must turn to the philosopher or literary scholar, who takes a few ideas and examines them in a hierarchy of contexts. He works with a vertical model that has more in common with Dante's theory of polysemous meaning in his letter to Can Grande than with the linear vectors of Descartes and Newton. Both Allan Bloom and Leo Strauss distinguish between secret and overt meanings in Socrates. John Keble's *Oxford Lectures on Poetry* posit a similar contrast between overt and hidden meanings in Virgil and Lucretius. Like Dante and Keble, most profound literature from the Bible and *Beowulf* down to *Paradise Lost* and Tennyson's 'Ulysses' opens the door on a retreating visionary world behind the natural world and on receding stories inside the stories, whose margins fade for ever and for ever as we read.

In the lottery of higher learning it may be better to divine the future like Virgil's soothsayers than predict events statistically. Instead of putting their trust in actuarial charts or laws of probability, scholars who are hungry for omens may have to 'haruspicate or scry' (T.S. Eliot, 'The Dry Salvages,' 5. 3–6) by defying 'augury not a whit.' As masters of multi-layered meaning, profound thinkers tend to decipher the future like Virgil, whose favourite genre is the omen poem and whose master trope is synecdoche. For as one commentator says, 'to believe that Virgil's lottery will yield a relevant truth, you must already feel that the source book, whether *Aeneid* or Bible, is so seamless a fabric that it holds wisdom whole' (O'Brien, 68). Even when surrounded by silence, an omen in Virgil is a lifeline that connects the interpreter to history because it is part of 'the world's tangled and hieroglyphic' script (Hart, xi). Unlike a computer printout, an augury or an oracle can properly be called a fragment because it is a hiding place of power, part of a concealed but greater whole.

Though most scholarly books are neither original nor profound, a few of them are original but not deep; or alternatively, deep but unoriginal. J.S. Mill makes an original contribution to ethics in his book *Utilitarianism*. And he lays the groundwork of induction in his *System of Logic*. But as F.H. Bradley shows in his attack on Mill in both *Ethical Studies* and *The Principles of Logic*, Mill's uncritical use of the pleasure principle and of associationist psychology can easily be demolished. His piecemeal methods are sometimes the opposite of profound. Conversely, a deep book may be profoundly unoriginal. I find H.L. Mansel's Bampton lecture, *The Limits of Religious Thought*, genuinely profound in its unorthodox use of agnostic theology. But in urging believers to take refuge in faith by exposing the impotence of reason, Mansel is merely fine-tuning a technique Kant had used with greater cogency and skill in *The Critique of Pure Reason*, a masterpiece that is both original and deep.

Often a contemplative scholar contributes most to the university by talking informally to students and to colleagues, by conducting tutorials on a voluntary basis, and by pursuing scholarship in a field outside his specialty. Long before Freud, Keble discovers laws of displacement in psychology and literary criticism by transferring the theological doctrine of reserve, a favourite topic of the Tractarians, to a new subject matter. It is difficult to predict how breakthroughs will occur. Abraham Flexner quotes one don as saying: 'I spent my life at Cambridge paid to do one thing and doing another' (267). Scholars have to be free to achieve less than they are expected to if they hope to achieve more. As Flexner says, 'the same conditions that permit idleness, neglect, or perfunctory performance of duty are necessary to the highest exertions of human intelligence' (267). Unless the university cultivates a tolerance for such anomalies, it is in danger of exchanging the diligent leisure that is essential to creative scholarship for a crude torture rack or traction device. The seminar table will then have less in common with the feast of discourse at Socrates' symposium than with the bed of Procrustes, a mere operating table for removing the cancer of heterodox opinion and for amputating limbs.

Many scholars spend their strength in channels of controversy or intrigue that have no great issue and are soon forgotten. But beyond politics and partisan debate there occasionally looms the prophetic voice of a J.H. Newman or Northrop Frye. As Frank Turner says of Newman, 'his gift was to transform his concerns and passions over the particular into a larger vision to which he virtually always appealed for his own limited polemical purposes, but which as the memories of the polemic faded left the universal categories of his message standing' (640). I have heard Northrop Frye transform and enlarge – as well as suddenly terminate – a polemic with a single aphorism or witty aside. Unfortunately, the petty rivalries or disputes that sometimes initiate debate are incommensurable with the mental agility and imaginative power displayed by every great scholar. Even when engaging in scathing, often *ad hominem,* attacks upon his philosophical inferiors, including so distinguished a cultural critic as Matthew Arnold, F.H. Bradley continues to tower and soar. But stinging assaults and withering wit sort oddly with a scholar's magnanimity. Though they may be consistent with mental rigour, they defeat flexibility, destroy friendships, and are seldom necessary.

Though someone once quipped that Isaiah Berlin received the Order of Merit for conversation, I doubt if a talent for common-room conversation is a permanent attribute of any scholar. By the conversation of the learned I mean something more intimate than the epilepsies of wit that convulse a high table exchange, the overflow of animal spirits in a senior common room, or even the sallies of a gifted raconteur. As soon as I forsake diplomacy and tact for rough courage in this matter, I find myself agreeing with Emerson that group discussions in seminars and common rooms destroy 'the high freedom of great conversation, which requires an absolute running of two souls into one' (1908d, 155). My mentor must be my friend, someone who exercises not only my intelligence and imagination but also me. Only two informed soul mates like Coleridge and Wordsworth or Emerson and Hawthorne, collectors of ideas who show each other their cabinets, can take part fully in the conversation of the learned. They must be in evanescent relation to each other, communing one to one, in the practice and consummation of a friendship.

Professors of humane learning are the rabbis and seers of a secular culture. As priests and Levites, the scholars preserve and transmit their culture's myths and liturgies. And as prophets, they subvert secular theology by using education as a low-voltage transformer to energize society with all that is wild and untamable in major art and literature. Moses said that 'all the Lord's people' should be 'prophets' (Numbers 11:29). Geoffrey Hartman believes that the attempt 'to honor yet control that prophetic energy, that speaking in tongues,' transforms all criticism into 'social criticism' (1980, 183). By making humane learning liberal, it also provides a welcome release of energy.

Even when we are moved by a great prophet or teacher, and by anything from a prayer, to the Gettysburg address, to a song by Bob Dylan, we can seldom explain its power. All we can say is that the aphorisms of Jesus' Sermon on the Mount have more in common with the metaphors in a sonnet by Shakespeare or with the paradoxes in Shaw's *Revolution-ist's Handbook* than with Euclid's theorems or a proof that God exists. Like the meaning of a joke or a pun, the aphorisms of a scientific reformer like Bacon, a prophet like Nietzsche, or a great teacher like Northrop Frye must be understood in a flash – all at once or not at all. When Nietzsche boasts that every aphorism is a peak and that it takes a giant mountaineer to leap from one peak to another, he is flirting like all prophets with danger. Yet as Stephen Booth points out, daredevils like Nietzsche and Blake also 'let you know they are married forever to particular, reliable order and purpose' (35). The boldness of the rebel is inseparable from the prophet's disciplined imagination and poetic vision. Frye says he has spent 'the better part of seventy-eight years writing out the implications of insights that have taken up considerably less than an hour of all those years' (1991, 55) When the humanist becomes a seer and when the scholar who has mastered his models becomes a prophet, 'the future is already here,' as Frye says, 'and the goal is now an enlarged sense of the present moment' (1991, 55).

5

Contemplative Knowledge:
A Secret Discipline

Though I argued in chapter 3 that scientists have more in common with scholars than with technicians, I now want to explore some equally important differences. For discovery and contemplation are as unique as Newton and Shakespeare. To talk with the great prophets, scholars, and poets of the past about justice, God, or Shakespeare's art is a much more erotic form of discourse than talking about quarks or genes, because (as George Grant has said) 'what is to be known about justice or God or beauty can only be known when they are loved.' Scholarly discourse may incorporate research into classical philology or numismatics, not because the great classical scholar is first and foremost a specialist on languages or coins, but because a scholar like C.N. Cochrane, the author of *Christianity and Classical Culture*, 'was first' (in Grant's words) 'an educated man who looked at the ancient world from out of his long life of sustained self-education' (1980, 7).

When offices of research administration try to assimilate all scholarship to science, they are making a serious mistake about categories. For though research is the precious lifeblood of science, the soul of scholarly discourse is the sustained (and sustaining) meditation of prophets, rebels, and poets. The scholar combines the bold surmise of a prophet with a rebel's impulse to criticize and subvert. His ascetic discipline and patience may also unite a monk's love of solitude and contemplation with a poet's insight. The contemplative discourse of the scholar includes the one-sided conversations students have with their instructors in their essays or that a professor conducts with Plato, Marx, or a still unborn posterity in a prophetic essay or reflective book.

The 'great library' venerated by F.H. Bradley's scholar as 'a temple of immortal spirits' may strike the technocrat 'as a most melancholy charnel-house of souls' (1930, aphorism 51). Like Francis Bacon, a scientist may ridicule the contemplative scholar's monuments for standing still 'like statues, worshipped and celebrated, but not moved or advanced.' Derided by technicians, however, a modern scholar has every right to invoke Bacon's own defence before a jury of sophists: he 'cannot fairly be asked,' as Bacon says, 'to abide by the decision of a tribunal which is itself on trial' (429–30, 438). Whereas the scientist or technocrat who hesitates to forget the founders of his discipline is lost, scholars often hold their best conversations with the immortal dead. Like the youth in the movie *The Sixth Sense*, they may communicate better with a society of dead poets and philosophers than with living friends. Without historians, there would be no national memories. And without the memories that a nation agrees to forget or remember, it would lack any shared identity. As individuals and nations, we commemorate what we care for and care for what we remember.

Though all knowledge is personal, to the highly individualized reader of a poem, science appears to express a truth no one knew before in language that even the most anonymous inquirer can know and understand. By contrast, to the anonymous knower of a science, poetry seems to express what everyone knew before in language too individualized for the knower of mere objective truth to decipher. Whereas a scientific knower is in principle identical with every other knower, poetic knowledge is unique to each interpreter. The scholar's world, like the poet's, does not authentically exist until it is brought into being by humanists who are by birth and by nature artists, architects and fashioners of worlds. Unlike the scientist, who has little interest in the ruined remains of Ptolemy's astronomy, the scholar is a connoisseur of ruins. He knows that civilization is itself a ruin which stands like Troy on the site of other ruins.

The forbidden knowledge that is half-revealed to Wordsworth and half-concealed from him in the great 'spots of time' passages in *The Prelude* is the godlike status of the contemplative poet's own imagination. The Romantic doctrine that god is the eternal self and worship of god is self-development turns the creator into both a marvel and a monster. For as a surrogate god, the creator may be a Frankenstein or Dorian Gray, an artist of potential immorality and crime. Though Wordsworth claims that the human mind is 'a thousand times more beautiful than the earth' because it is 'of quality and fabric more divine' (*The Prelude*, 14.449, 454), he also feels alone in the universe, isolated and unsponsored, or (as Newman prefers to say) 'without God in the world' (1968,

187). As an exiled parcel of divinity, he is an orphan and rebellious out-cast, stranded like Wallace Stevens's aesthete 'in an old chaos of the sun' ('Sunday Morning,' 8.5).

In *The Logic of the Humanities*, Ernst Cassirer speculates that we love to receive or impart wisdom, because wisdom is something added to knowledge that we ourselves contribute. Wordsworth's 'spots of time' in *The Prelude* are more terrifying than consoling, partly because he tries to repress the truth that the logic of the humanities is a logic of self-creation. As architects of Yeats's Byzantium, the city of art and culture, humanists are in love with the social self they construct, partly because this second self supplements a natural self bestowed on them at birth, a self they may neither like nor understand.

It is in the interest of at least some scholars to be learned amateurs or liberally educated specialists. Because great scholarship is unregimented and speculative, it will always be easier to divide watchers of the skies into astronomers and astrologers than to separate philosophers or literary critics into wise men and cranks. To know only one's own discipline is to be ignorant even of that. Perhaps only a close encounter with philosophy or history will liberate a literary scholar into the vividness of a new idea or the ardour of a passionate lecture or prophetic book. Jaroslav Pelikan shrewdly observes that a specialized graduate education will determine the difference between a good and a bad doctoral dissertation. But even more discerning is his remark that only a broad liberal education will mark the difference between a good book and a great one.

The precept that everything is what it is and not something else sup-ports Isaiah Berlin's wry conclusion that some great goods cannot exist together. Essays on philosophy, history, and literary criticism are *sui generis*. Since each belongs to a distinct genre of writing, it may confuse a scholar to mix them. Though disciplinary purity breeds death, A.N. Whitehead believes that the only way to avoid mental atrophy and decay is to teach thoroughly a few seminal subjects (1933, 14). We love what is specialized and personal. 'Wherever you exclude specialism,' warns Whitehead, 'you destroy life' (1933, 22). Moreover, it is always easier to be fashionable than farsighted. No scholars have a monopoly on intel-lectual fraudulence. But passing as authorities who invite that 'mixture of disdain and amusement that professionals reserve for amateurs' (Fleishman, 132), scholars may teeter on the edge of pretension and imposture. If they stray too far from their fields of competence, they may even lose their balance and sink in the quicksand of pop culture and pornography.

It is better for a scholar to go a long distance on only one path than a little way on many. Whereas the media guru knows nothing about everything, a specialist's knowledge is like 'the space between two parallel lines.' As J.S. Mill says of Bentham's labours, this space is 'narrow to excess in one direction,' but 'in another it [reaches] to infinity' (1950, 75). A specialist who occupies a fixed centre can use the arms of a compass to sweep out the lines of a large parabola. By contrast, a scholar who poses as an expert on everything drifts in space like a lost astronaut. He is stranded on an arc that has no centre.

I once knew a classical philologist who prided himself on his knowledge of literary theory. All the literary theorists I talked to felt he was too conservative to be a theorist. But his colleagues in the classics department were equally convinced that by tilting at theoretical windmills he had become too quixotic to be a respectable philologist. The best scholarship combines the knowledge of a specialist with the invention of an amateur. It is imaginative but informed. Studies of culture that suppress critical intelligence and responsiveness to language are often original but seldom authoritative. They combine the pedantry of a specialist with the knowledge of an amateur. In the philologist who pontificates on pornography or the linguist who writes books on psychoanalysis and the history of sex, we witness a drizzle of scholarship dissipating in the air.

Indeed, even the most innocent effort of mathematicians and literary scholars to break out of their specialties may become a sortie fraught with risk. As Northrop Frye amusingly recounts, 'I thought that exclamation marks in the equations of the mathematicians represented some enthusiasm for the beauty of their subject, and that there was at least that much communication possible; but I was undeceived on this point.' Though mathematics and literature are both in search of an audience, Frye wryly concludes that 'any such subject as "communications," designed to communicate better between them, tends to become enclosed in its own special language, and merely to add one more unintelligible voice' (1988, 66).

My high hopes but low expectations for interdisciplinary studies were confirmed several years ago when I foolishly agreed to take part in a comparative literature colloquium on literature and the visual arts at a western university. The best part of the event was a civilized conversation over sherry with the two other panellists in a small library adjoining the conference room. Unfortunately, in the main room itself some kind of thermal inversion had made the atmosphere convection-free: in a crowded hall that had about as much oxygen content as a mountain peak in Nepal, the audience was already panting, not for revelations, but for air. I, too, felt short of breath when the first speaker, an expert on surrealism in French culture, and a master of suave and elegant English, chose to talk entirely in Italian. As melodious vowels rolled off his tongue, I felt my head nodding. Despite my best efforts to keep awake by biting my lower lip and drinking more coffee, I lurched back to consciousness only when a wag at the back of the hall directed a question to me about the bracing discourse we had just heard on art, neurosis, and clinical depression.

The second panellist, a towering personage who spoke in French, used a microphone adjusted for the first speaker, who was eight inches shorter. Though the microphone amplified the rustling sound of her notes, which she kept shuffling on the lectern, it did nothing to project her high-pitched voice to the audience. I recovered a small portion of my flagging energy when I saw to my alarm that she was analysing slides of Dürer's *Melancholia* that I was going to compare with James Thomson's portrait of Melancholia in *The City of Dreadful Night.* When the French scholar finally asked the projection booth if this was her last slide, I had to tell her she was already three slides into my own talk. To my City then I came, burning, burning, burning, with the snicker of suppressed laughter smouldering in my ears.

Collingwood thinks that to shine at learned conferences 'one should have a rather obtuse, insensitive mind and a ready tongue' (1939, 54). Since I tend to chatter at interdisciplinary conferences with a fluency directly proportional to my ignorance, I now limit myself to conferences on Victorian poetry, where I hide my resemblance to a parrot by thinking more and talking less. Despite the traps that lie in wait at an interdisciplinary conference, a contemplative scholarship faithful to its own vocation has to foster knowledge that combats disconnection. We live in a specious present of transitory e-mail messages, discontinuous TV news, and spot commercials. The longer the line-ups in college libraries to use the computer terminals, the better the library may think it has achieved its inhuman goal of becoming a factory or assembly line. To counter such forces the university must recognize that genuine learning always involves time set aside for solitude. By contemplating a sorrow and a calmness all its own, a solitary soul may magnify its life. But to discover a 'majesty of grief' that may even allow us to 'enhance and glorify this earth' (Arnold, 'To a Gipsy-Child by the Sea-Shore, ll. 20, 68), we need to go inside ourselves and cultivate 'great inner solitude' (Rilke,

54). Rilke says that only the universe we contemplate out of the breadth of our own aloneness can become our true work, rank, and career.

Whereas Rilke believes in the wise incomprehension of the Romantic child, whose aloneness puts him in touch with deep cosmic laws, Thomas Merton's model in *Entering the Silence* is 'infused contemplation.' The scholar stands in the same relation to the wisdom or knowledge he chooses to contemplate as Merton's priest stands to Christ. Just as the priest becomes 'one of those masks behind whom Christ hides and acts' (Merton, 164), so the scholar must cultivate behind his public persona the private wisdom of the inquirer who approaches truth in the darkness of his ignorance and who is most knowing in the way he does *not* know. Like Merton's priest, the scholar is often 'afraid of being absorbed in the public anonymity' of his vocation. 'I think of so many priests I know in their strange, sensitive isolation,' Merton muses, 'innocent, hearty men, decent and unoriginal and generally unperplexed, too; but all of them lost in a public privacy' (164–5).

The oxymoron of a public privacy means that, as lecturers and writers, scholars are 'everybody's property,' even though as friends of contemplation and solitude they belong to wisdom alone. Everything transparent and transmissible in a scholar's lecture will be memorized by the dullest students and reproduced on their exams. But if a scholar should take the risk of composing himself to silence and opening his mind to wisdom, many students will complain that his discourse is disorganized and his ideas obscure. Every scholar worthy of the name must try to find a few moments in each lecture when he dies on purpose to the rush of time. But to survive as a teacher, he must also pay the price of his strange vocation. For in his 'public privacy' he will probably reach only a handful of students.

Too often scholars and scientists who dedicate themselves to intellectual enlightenment fall 'into the great indignity' that Merton warns against: like his Cistercian monk, they are 'contemplatives' with no time to contemplate, who seem always 'ready to collapse from overwork' and failure (381). Their best conversations are seldom between themselves and their colleagues or students but between themselves and the immortal dead or even God. 'There is so little one can communicate,' Merton complains of the novices. 'I talk my head off and they seem to be listening but, when I ask them questions, I find they have been listening to somebody who wasn't there, to stories I never told them' (381). In a search for enlightenment, scholars and monks suffer a double indignity: they sacrifice contemplation to the gross busy-ness of their efforts to instruct overworked or indifferent novices; and then they fall into the further error and 'indignity of thinking such an endeavor is really important' (381).

Unlike the monk, who belongs wholly to God, no scholar belongs wholly to his discipline. Though he should try to replace the suspicion of rumour-mongers with friendship or trust, it must be a trust of all students and colleagues, a 'love for everybody' as Merton explains (382), 'equal, neutral and clean.' But to 'love everybody' is to be 'possessed by nobody,' and so to be a version of Keats's non-moral chameleon, a paradox of negative capability. As someone who is 'not held' or 'bound,' and whom 'nobody' therefore 'knows' (Merton, 382), such a person must give away his soul and become the opposite of a human personality.

Scholars, like monks, must accept the responsibility of living with other people while avoiding 'the crass activism that delights in company and noise and movement and escapes the responsibility of living at peace' with the enlightenment they seek (Merton, 389). It is also important not to mistake the means for the end: Merton is so consumed by an obsession to correct the pitch and tempo of the monastery's choir that his duties as sub-cantor become a torment to him. When commencement addresses and the conduct of public relations with alumni magazines divert attention from the scholar's true mission, they produce a degenerate 'worship' of 'machinery,' the phrase Matthew Arnold uses to denounce the idolatry of means. Our 'whole life must be a dialectic between community and solitude,' says Merton. Conversation enriches the understanding, but solitude is Merton's Muse, the school of his genius. 'Both are tremendously important,' and the 'contemplative life' of the scholar, not less than the life of the monk, 'subsists in the fruitful antagonism' between the two needs (389).

When malice and suspicion kill the soul of an intellectual community, ordinary friendship and love are seldom disinterested enough to restore peace. Though Merton trembles when he meets the novices because he is afraid of being 'immolated in their presence' (397), such a sacrifice entails the killing of 'what is not right in his love' for them. For as an effective teacher he is 'bound to love' his students 'without being in love with them' (397). For Merton the paradox of a teacher's ability to love without loving, and to care without caring, is resolved in the mystery of the Mass, 'where we are all alone together and ... are offered to God together' (397). Like Merton, all dedicated teachers make a sacrifice of their imagination and intellect, giving to the scholars who come after them an essential portion of their own lives. Just as 'everything is related to God but God is related to nothing' (Merton, 399), so scholars and teachers must cultivate in themselves and their communities an affection for the wisdom that can be shared but not communicated in words. They must love people 'for what they are, not for what they say' (Merton, 398).

By an odd paradox the most practical habits are also the most contemplative, and vice versa. A monk who contemplates a God of love should have some practical experience of love before contemplating him. But until contemplation of God refines or sublimates a monk's experience of love, how can he contemplate God at all? A similar dilemma confronts the literary scholar. She should have some experience of the well-commingled images and sounds in Robert Frost's poem 'Spring Pools' before she tries to explain what may be too implicit to make explicit, too perfect (as it were) to 'explicate' or explain in words. But until the literary scholar contemplates the whole poem, how can she meaningfully experience or try to explain any component part of it? The dignity of literary criticism and scholarship comes, not from the fact that critics teach poetry and write books about it, but from the fact that these activities derive from contemplation and from an opportunity to be reflective (if not be wise) about words. Scholars who read widely without contemplating what they read are seldom original thinkers. 'You should read only when your own thoughts dry up,' warns Schopenhauer; 'to banish your own thoughts so as to take up a book is a sin against the Holy Ghost; it is like deserting untrammelled nature to look at a herbarium or engravings of landscapes' (90). If we try to contemplate literature or God without some practical experience of their component parts, and if we try to understand the parts without the whole, it is difficult to see how either contemplation or practice can begin.

Even literary scholars who value poetry and criticism as contemplative activities may disagree about how critics may experience the purity of art without violating what they contemplate. Geoffrey Hartman is shocked when his effort to pay 'intellectual homage' to poets by writing creative criticism is attacked in the 1970s by a senior colleague at Yale, William K. Wimsatt, who accuses Hartman of 'battering the [literary] object' (Hartman, 1999, xxi). Ironically, Hartman's criticism of cultural studies two decades later is uncannily similar to Wimsatt's earlier attack on Hartman. In blaming cultural critics for causing 'the disciplinary line between social study and the study of art [to] practically disappear' (1999, xxviii), Hartman repeats Wimsatt's charge that there can be no substitute for an original contemplative criticism, which should be admired 'for its daring, eccentric orbiting of the text and for its integrity as prose' (1999, xxv).

Poetry may be obscure, not because the thought is difficult, but because it is largely non-existent. I am thinking of Symbolist poems by Baudelaire or Wallace Stevens, which are too pure for ideas of any kind to violate. Since great literature asks us to look *at* language as well as *through* it, its meanings are often 'constitutionally opaque.' They may even be 'inherently mysterious' and 'undecidable' (Donoghue, 1997, 131). To assume otherwise is to destroy literary criticism as a branch of higher learning by assimilating its models to those of some currently ascendant discipline, usually one of the social sciences. In an age of speed reading, only literature that is pondered with a fond and lingering scrutiny can foster delight in words for their own sake. It is as natural for literary critics to contemplate words and for scientists to discover physical laws as it is for Hopkins's just man 'to justice.' By contemplation and discovery I refer not merely to interior states (as when someone is said to be in a state of ecstasy or shock) and not merely to the official vocational obligation of a scholar or scientist to his academic community. By the literary critic's contemplative bent or the scientist's passion for discovery, I refer to something more interior than a professional rank or office but to something less wayward or fugitive than a mere predilection or whim.

I suspect that it is as hard for a contemplative scholar to be a media guru like Marshall McLuhan as it is for a rich man to enter the kingdom of heaven. On the other hand, I have known scholars who use a solitary life, not to temper the sword of a keen mind or enhance their response to great art and thought, but as a narcotic to dull the pain of unfulfilled promise. Even a monk may embrace solitude, not primarily to contemplate God, but to sever the goodly fellowship of dust before it is timely. Yet the abuse of a contemplative life, like the abuse of an active one, is no argument against its proper use in contemplative scholars like Newman and Merton, who share a genius for solitude. Even when Newman was 'very much alone' at Oriel College, he was not lonely, as Dr Copleston recognized, because solitude was the atmosphere in which his mind and soul breathed. In his *Apologia* Newman recalls that Copleston 'turned round, and with the kind courteousness which sat so well on him, made a bow and said, "Numquam minus solus, quam cum solus"' (1968, 25). Since a scholar's calling is to arouse doubt and train students to dissent, their departure, like death, is always at hand and should be expected when it comes. A great master encourages disciples to reject his teaching. As Newman discovers during his last months at Oxford, his vocation, at the end, is to be alone.

Scholars of genius are often absent in their small talk because they are present somewhere else. As someone who tried unsuccessfully for twenty years to make conversation with Northrop Frye at the Victoria College high table, I was less surprised by his kindly but remote treatment of me than by the talent with which he used his duties as chancellor as an opportunity to put into play his deepest convictions and liveliest wit. Like most creative people, contemplative scholars who value their solitude often seem testy, angular, aloof, or even cold. As Timothy Findley says, 'whatever form the genius takes, there are signals given ... Some of these can be off-putting at first. For instance, a sense of inviolable aloofness (think of Baryshnikov); a private, often eccentric vocabulary (think of Marshall McLuhan); coolness (think of Meryl Streep)' (275).

Everything in a scholar or scientist may cry out for the discipline of study, the rigour of controlled experiment, or the daily habit of reflecting in solitude. But the potential scholar may be excluded from the charmed circle by the need to serve a nation, a city, or even so pedestrian a body as a college committee. Whereas David Damrosch wants the scholar to have communal identity, David Bromwich calls for radical solitude and isolation. As Merton's journals make clear, community and solitude exist in uneasy tension, even in a monastery. Preference for solitude is not an act of rebellion against community but the condition of a scholar's being reflective and productive. As the monk of a secular age, the scholar who hopes to receive 'the spark from heaven' and then impart a portion of that wisdom to the world may first have to go into seclusion like the Desert Fathers or like Matthew Arnold's Oxford student in 'The Scholar-Gipsy.'

It is always easier to say what contemplative scholarship is *not* than to say precisely what it *is*. Clearly, it is not a craft reducible to teachable rules. Nor is it a science used to predict or control events. As an art rather than a science or a craft, scholarship seems to contribute to the civilized conversation of mankind by combining features of three activities: play, games, and purposeful 'experiments of living,' a felicitous phrase of J.S. Mill. There are no rules for producing great scholarship or for arriving at a state of secular grace, just as there are no rules for having insights or for producing great poems or paintings.

Contemplative scholarship is best pursued in that spirit of play which Whitehead finds 'energizing as the poet of our dreams' (1949, 97). It should also be conducted in a spirit of competitive gamesmanship. As the architect of our purposes, the scholar stands in the same relation to an organized body of knowledge as God stands to the laws of a rational universe or as an aeronautical engineer stands to his spaceship. Just as a purposive education can be debased into mere technical training, devoid of any goals or models, and just as the play principle in scholarship can be trivialized as the pursuit of what is merely frivolous, so the competitive game spirit that is creatively harnessed in politics and law can break loose into what Bacon calls contentious learning, a form of intellectual warfare marred by self-defeating quarrels and disputes.

Distinguishing among three kinds of education (liberal, vocational, and aesthetic), O.B. Hardison argues that the third kind should replace the first. Liberal education is based on classical and Renaissance models of learning that were practical rather than theoretical and that once promoted the ideology of a ruling class. Rhetoric and the imitation of classical models produced a virtuous man, and they freed him to defend himself by persuading judges in law courts and the forum. But liberal education has become increasingly contemplative and aesthetic. By cultivating art for its own sake, a sense of destiny with no immediate destination, it has restored faith in the Romantic doctrine that scholars are builders as well as archivists, creators as well as custodians of the Word.

Like many defenders of humane learning, Northrop Frye has difficulty identifying within the liberal Socratic tradition a convincing alternative to Newman's gentleman scholar or to Arnold's cultivated member of an elite social group that makes its humanity rather than its class traits prevail. According to Hardison, Frye's perplexity arises from his confusion of two traditions – the liberal and the aesthetic. Frye persists in honouring and reinventing Socratic humanism, including the Socratic dialogue, even though he himself was the least Socratic of teachers – an oracle rather than a midwife. Hardison pokes fun at the hypocrisy of the modern Socrates, whose method is a mere charade: 'Trying to maintain the Socratic posture leads teachers to coercion and intellectual hypocrisy' (137). I think, however, that the Socratic method can be internalized in intellectually alive lectures that retain the best features of the liberal model. Socrates is given a lease on life in an aesthetic tradition best represented by Kant, Schiller, Cassirer, and even by Frye himself, despite the fact that in his oracular or prophetic moments Frye seems to be borrowing the methods of an alien liberal tradition.

Frye boldly argues that literary criticism and the study of language occupy the same place of privilege in the scholarly world that mathematics occupies in science. It is a foundation study, not merely one discipline among many. But the great antidote against the prostitution of literature departments into purveyors of technical writing skills is provided, not by Frye himself, but by Richard Lanham, who shrewdly deploys utilitarianism's pleasure principle against utilitarianism's most vocal advocates. If utility is to be the measure of competent writing, it is important to recall that utilitarianism is designed to promote the greatest happiness of the greatest number. Pleasure rather than clarity is the criterion to be invoked in any utilitarian theory of value.

As an advocate of playful impersonation and role-playing in higher learning, Lanham is most engaging when conducting his spirited polemic against 'the gospel of normative clarity' (1974, 137). Deploring theories of style that 'pare away all sense of verbal play' and of a 'self-satisfying joy in language,' Lanham blames computer technologies for masticating our delight in language 'into binary bits' (1974, 10). As the science of language, linguistics is also the great enemy of verbal ornament and play. Unlike linguistics, which woos mathematics, literary criticism should make love to eloquence and take pleasure in verbal imitation for its own sake (1974, 67).

The soul of higher learning is that behavioural wild-card ethologists call 'vacuum activity' – the desire of ducks to upend themselves to feed under water, not because they have to, but because they enjoy upending for its own sake (Lanham, 1983, 44). Spontaneous activity or play, the often risky impulse to do something that is 'altogether beyond purposive explanation' (Lanham, 1983, 133), means that there is nothing outside liberal education, nothing more valuable or dignified, to justify it.

Instead of instilling a particular ideology or imparting economically useful skills, contemplative scholarship affirms that 'the central value of education' is 'the freedom of the individual and the enlargement and enrichment of his inner life' (Hardison, 17). Such education achieves these ends by liberating the student's imagination and encouraging a free play of the mind. Frye believes that 'the genuine energy of the arts and sciences converges on a world where work and play have become the same thing. A gathering together of such people with such interests ... would be in the deepest and most serious sense a play *ground*' (1988, 167).

The humanist's free play of mind has three defining qualities that it shares with all higher learning. In the first place, it expends energy for its own sake, unlike work (which is energy expended with a further end in view). Second, free play is undertaken in a spirit of enjoyment and even love. It is an antidote to the brute coercion of the sophist, who browbeats his adversaries and argues for victory rather than enlightenment. Coercive play is to free play what rape is to pleasurable sex. Or in Simone Weil's loftier phrase, the mental play of a great thinker is 'intelligence illuminated by love.' Finally, play postpones closure by deriving as much pleasure from the journey as from the destination itself. Like scepticism, play has the strength to remain in uncertainties. In his most playful dialogues Socrates dramatizes the blessing of a learned adversary, 'not just any enemy,' as David Damrosch explains, 'but someone very like yourself in every respect except for a pigheaded insistence' on being wrong on one point (83).

The deepest purpose of a game is the delight it provides in being competitive, in honing one's mental or physical skills, and even in enjoying the challenge of taking part in some activity for its own sake. The game of finding a cure for diabetes before one's rivals or of beating a colleague in a contest for a fellowship is a competitive form of play. Though in one sense it is important to win a game, in a deeper sense it is not. The ostensible purpose of golf is to put the ball in the cup. But unlike a life-saving medical procedure or a defence attorney's summation to a jury in a murder trial, the purpose of sinking a putt is oddly purposeless. Like a Socratic dialogue or the pursuit of virtue, the medical scientist's competitive investigations into the genetic origins of disease or the philosopher's delight in the intricacies of a moral argument can confidently be left to justify themselves.

Frye says 'a life divided only between dull work and distracted play is not life but essentially a mere waiting for death' (1967, 63). Playful work or meaningful play is a privilege scholars and scientists gratefully enjoy rather than a duty they grudgingly perform. Though researchers may invent pragmatic arguments to justify their grants, their gamesmanship and play cannot be rationalized. Scholars who value financial remuneration and victory over their competitors more highly than the free play of the mind in the pursuit of truth are often more outspoken or 'visible' than their counterparts. But their specious defence of vocational or professional training is always far removed from the demanding but liberating high adventure of a scholarly vocation, which is a life of continuous self-creation that no higher principle has to justify.

Though Matthew Arnold insists that culture is the 'humanisation of man in society,' he recognizes that self-growth is often at odds with activities society calls useful. As one critic says, 'if the social order were

perfected, we would all be reduced to the level of the bee or the ant'
(Redfield, 161). In opposing the mechanical, anti-creative routines that
are the god of every social system, 'the Socratic teacher must take care
that his teaching does not degenerate into just another method. If he
makes his students his disciples he has become a sophist, and has failed'
(Redfield, 162). Since sophists make good clones but poor inventors,
the modern Socrates is most faithful to the master's example when he
ceases to be a disciple. Paradoxically, he is most obedient to the master
when rebellious enough to disobey him.

Like behaviour for its own sake, which is the mysterious 'wild-card' of
higher learning, the sportiveness of a game seems 'to lie outside the
purposive paradigm' (Lanham, 1983, 133). But there is an important
difference between the games of discovery that energize a scientist and
the free play of the mind that liberates a scholar. As a game to control
and predict events in the physical world, science is more competitive
than scholarship. It inhabits a space halfway between the 'great land of
Spontaneity' or Play and the public world of engineers and lawyers,
where everything is justified by practice or utility. Sometimes it is difficult to differentiate among play, game, and purposive activity. Though
scientific games can be purposive, their competitive urge to defeat rival
theories and conquer ignorance may be stronger than the impulse to be
sportive. As Lanham explains, 'Huizinga vacillated about' the importance of purpose and play. 'Sometimes he could find purposes, sometimes not – and his predicament ... brings into focus humanism's central
paradox. If play and game form a new behavioral paradigm, a conception of motive inexplicable to purpose, musn't we give up and simply
contemplate these activities' as examples of disinterested behavior, or as
behavior for its own sake (1983, 133)?

A book on higher learning that is less playful or self-critical than the
learning it advocates invites a sermon against sermonizing. But by venting a partisan intensity of distaste, even a homily against moralizing may
drug the mind by substituting a dead dogma for a living truth. Serious
play, or what Elizabeth Sewell calls the 'field of nonsense,' is best displayed in a book like Stephen Booth's remarkable monograph *Precious
Nonsense*. Booth shows how great works of literature like Ben Jonson's
epitaphs or Lincoln's Gettysburg Address often manage to express more
than their authors consciously *know* they know by combining folly in one
direction with wisdom in another. Though Nietzsche condemns the
Beatitudes as a conspiracy of the sheep to convince the wolves that it is
sinful to be strong, the precious nonsense of the meek inheriting the

earth or of the poor in spirit possessing the kingdom of heaven is also a
magnificent celebration of spiritual emancipation. Looked at in one
way, their replacement of Mosaic law with a law more subtle and exacting is an act of sublime folly. But looked at another way, the Beatitudes
are a liberal education of the spirit that replaces the law of moral command – 'thou shalt not kill' – with a hymn to the holiness of the heart's
affections and the truth of a purified imagination.

Oracles may be discrete fragments or atoms of discourse, as in F.H.
Bradley's book of aphorisms or Shaw's *Revolutionist's Handbook* in the
appendix to *Man and Superman*. Alternatively, they may proliferate like
fractals, generated by a describable law or rule. Hovering midway
between the fragment and the fractal is a discourse like the Beautitudes,
which seem at first glance to consist of discrete fragments, but which on
closer inspection turn out to be fractals, oracles that are as prescriptive
as the Ten Commandments, generated by a liberal law of their own. In
the Beautitudes peace-making, poverty of spirit, purity of heart, can be
justified by an explanatory 'for' clause: 'for they shall be called the children of God,' 'for theirs is the kingdom of heaven.' Alternatively, the
'for' clauses can be appositive rather than causal. They can be con-
strued, not as an explanation, but as a praise and celebration. For like
liberal education, purity of heart and poverty of spirit are self-justifying
affections in no need of explanation. Just when we assume, however,
that our purity of heart liberates us from the law, the sublime nonsense
of abolishing a law only to consummate or fulfil it may require us to
make our observance more rigorous than ever: 'Think not that I came
to destroy the law or the prophets: I came not to destroy but to fulfil'
(Matthew 5:17).

Equally sublime and evasive are the playful two-way meanings of the
Lord's Prayer. Is God remote or immanent? Is his kingdom of earth or
of heaven? Is divine behaviour a model of human behaviour, or is it the
other way round? Does God command his creatures, or is the Lord's
Prayer an order to God in disguise? Its familiar words seem to supply
opposite answers to these questions simultaneously. Definitions of God
as the being whose existence is always being proved or disproved, or
whose centre is everywhere and whose circumference is nowhere, make
inhuman sense. At the opposite pole is the precious nonsense of trying
to give God a local habitation and a name. To the extent that oracles are
local and human, they are precious. To the extent they reflect the divine
in the human they are hard for logic to digest and may seem more like
sublime nonsense than wisdom.

C.S. Lewis points out that even the first line of the prayer 'goes to pieces if you apply the literal meaning to it. How can anything but a sexual animal really be a father? How can it be in the sky?' (1966, 296) Compared with the philosopher's proposition that 'the supreme being transcends space and time,' the prayer's claim is 'precious nonsense.' But precisely because it *is* 'nonsense,' the language of religion, like the language of poetry, retains a strong trace of dream that allows it to represent a more concrete and therefore more precious experience than the mere abstraction defined by the philosopher's clever use of counters.

Donald Spoto explains that 'The Lord's Prayer begins with a reworking of a well-known Aramaic prayer, the Kaddish spoken at the end of the synagogue service: "Magnified and sanctified be His great name in the world which he has created according to His will ... May He establish His kingdom ... speedily and soon." Jesus turns the prayer just a bit: the kingdom becomes an event now, when God's will is done on earth as it is in heaven – that is, the kingdom here below (man's acceptance of grace) localizes and makes visible the kingdom' (172–3). Though God's kingdom is still to come, its advent can be hastened. Even now an eternal greatness can be brought down to earth. Like the speaker himself, it can be incarnated in natural and temporal fact.

> Our Father who art in heaven,
>   hallowed be thy name.
> Thy kingdom come,
>   Thy will be done on earth, as it is in heaven.
> Give us this day our daily bread;
>   and forgive us our trespasses,
>   as we forgive those who trespass against us;
> And lead us not into temptation,
>   but deliver us from evil.
> For Thine is the kingdom, the power and the glory,
>   for ever and ever. Amen.

Jesus' slight reworking of the Kaddish is important, since if God's kingdom comes, then the mystery of that coming will be an achieved mystery, a sublime fact rather than a distant hope. To the degree that the prayer's petitions compose a felicitous rite, they will become an example of 'doing by saying,' like God's creation by verbal fiat at the opening of Genesis.

'Give us this day our daily bread' is ostensibly a prayer. But if the Lord of life withholds this day our daily need, he is not a lord of life. Since the prayer implicitly contains its own desired response, it is an order to God in disguise. The chiasmic form – 'Give us this day our daily gift' – shapes the reasonableness of the prayer. How can a rational father deny so rational and reasonable a request?

Instead of making God's will a model of human volition, as in the opening prayer, God is now asked to forgive us our trespasses, because we ourselves have done no less for our enemies. 'Forgive us as we have forgiven (not *you*) but those who have trespassed against *us*.' A noble and generous *human* response is now appealed to as the model of God's treatment of his people. If God cannot forgive us, he will be less forgiving than his creatures. Could we forgive such a God? Or would such forgiveness not (in a sense) be the ultimate trespass against his own creation?

Does 'lead us not into temptation' mean 'lead us' into something other than temptation: lead us elsewhere? Or does it mean merely, 'do not lead us' into evil? Where is the caesura to be placed: after 'us' or after 'not'? The ambiguity repeats itself in the use of 'deliver.' The God of deliverance may take us into his Kingdom; or he may merely deliver us from bondage. 'Deliver us from evil' falls short of a full triumphal entry into power and glory.

The prayer seems to resolve all its equivocations in a final sequence of resounding 'for's: 'For Thine is the Kingdom, the power and the glory, for ever and ever.' The 'for' of the double 'forgive's ('Forgive us our trespasses as we forgive ...') is repeated in the climactic causal 'for' of 'For Thine is the Kingdom' and in the final 'for' of the repeated adverbs: 'For ever and ever.' But these phonetic similarities mask profound disparities of meaning. Is the conclusion a triumph of resolved ambiguity and logic or is it a trick of mere phonetic symmetry?

Great literature like the Lord's Prayer or the Beatitudes has less in common with a computer programmed to play chess than with a human chess master. Whereas a computer undertakes a heuristic search by rapidly examining a whole forest of possible strategies from which it lops away the branches of defective trees, a chess master has a subliminal or fringe consciousness of many trees. Unlike a computer, whose strength depends on its storage capacity, the intelligence of a scientist or poet is not seriously diminished by a partial loss of memory. A computer seeks context-free precision, but a poet thrives on a complex play of meanings. Though poetry and science are both passionately particular, poetry is as hostile to context-free particulars as science is indifferent to diversity.

Any preponderance of prophecy over history is a degree of insanity. When Jesus talks to his Father we say he is praying. But when Job says God speaks to him out of the whirlwind, or when the mourner claims that the breeze starts talking to him at the climax of *In Memoriam*, we may suspect that each speaker is less inspired than crazy. In illuminating and confirming the high drama of subverting authority to renew tradition, however, Jesus also shows how life can be lived at its limits. The sublime play of hints half guessed in the Lord's Prayer and the Beatitudes is so exalting that perhaps their prophecies should be accepted as true even if they are only noble follies – species of what Stephen Booth calls 'precious nonsense.' Like the Gospels' riddling parables, the Beatitudes keep us off-balance. Their oracles and prophecies elude our grasp even as we seem to understand them. Coleridge traces the etymology of 'aphorism' to the same Greek word as 'horizon,' the word that means 'bounded' and hence 'the limit of our vision' (17). My habit of returning to parables and aphorisms, always frustrated but often renewed, takes me as close as possible to Shakespeare's experience of the phoenix and the turtle – 'Single nature's double name / Neither two nor one was called' ('The Phoenix and the Turtle,' ll. 39–40) – and so to an experience of phoenix-like rebirth or resurrection.

For many scholars spiritual authority resides today, not in the churches or synagogues, but in the centres of higher learning. Unfortunately, even when society pays lip service to a university's hypothetical languages and hypothetical sciences, its temples occupy the same position as the Musical Banks in Butler's novel *Erewhon*. Everyone pretends to pay homage to these banks, but no one deposits any money there. While unwise power is concentrated in the business schools and governments, a powerless wisdom is the fate of most prophets and scholars.

To be welcomed for its lightness of spirit, the paradox of wise or noble folly is the great solvent of self-importance and pretension. It is also a reminder that the soul of higher learning is the pursuit of knowledge in a spirit of serious play. Alcibiades registers the playful, foolish side of Socrates by comparing him in the *Symposium* to an ugly Silenus or pugnosed satyr. And Erasmus uses the pagan female figure of Folly to celebrate Jesus as a wise fool who cures the 'folly of mankind' by choosing 'the foolishness of the Cross.' In acknowledging the spirit of play in wise fools like Socrates and Jesus, the academy resists Matthew Arnold's impulse to consecrate high seriousness by cloaking culture in a pall. Any Don Quixote who sets out like Henry Adams in search of genuine knowledge is likely to suffer the fate of the comic fanatic who finds his dreams escaping him simultaneously on all sides. The more order Adams seeks, the more disorder he finds: discipline dissolves into anarchy, unity into fragmentation and chaos. Tilting at windmills whose vanes keep turning on him, the modern Quixote may suffer the indignity of playing God's fool after God has disappeared and Socrates has been replaced by a microchip.

The hostility to contemplative scholarship is sometimes a resistance to the notion that the university is a place where professors love what they teach. Sophists wonder why talented professors should waste their time diverting energy from the contemplative scholarship they do supremely well to the teaching of undergraduates who seem to learn just as much from novices as from veterans. The answer is that to contemplate Mozart's music is to love it. And though love is often refined in solitude, to be deepened it must be shared. For this reason it is never absurd for St Paul to serve as a parish priest or for Michelangelo to teach art to a kindergarten class. Before assuming his duties as president of Harvard, Derek Bok was told by one alumnus that his first act in office should be to abolish Harvard College (Bok, 35). Bok rightly believes, however, that an important value of liberal education is the mingling of Nobel laureates with Harvard freshmen. The young apprentice is offered 'an authenticity, a depth of understanding, a subtlety of perception that comes from constantly trying to create new hypotheses and interpretations that must sustain the weight of informed criticism' (36). And sometimes, Bok thinks, the benefits are reciprocal. What better way for a senior scholar to write a first draft of a book than to make his 'knowledge intelligible to an audience that is interested but unprotected by the heavy armor of disciplinary technique' (36)? By leaving her archive or study for the seminar room or lecture hall, the celebrated critic or philosopher is able to make discursive thought presentational. She is able to embody what she knows in a performance.

In a society drowning in information but starving for knowledge, scholars must learn to work alone as well as in research groups and teams. Solitude is increasingly important the more a scholar knows. In college a third of the instruction should be self-education. In Newman's words, 'the conversation of all is a series of lectures to each' (quoted by Pelikan,

61). In graduate school self-education should occupy more than half the scholar's time. What is done alone in the library, the laboratory, or the study is more important than the conversations one shares with instructors or peers. Indeed, as a scholar becomes part of the great Burkean contract 'between those who are living, those who are dead, and those who are to be born,' she will be in closest touch with colleagues at other universities or even with scholars of a different generation.

Like wisdom, contemplation is not to be weighed in the scales of any doctrine or teaching. 'Indeed,' as Merton says, 'it is not to be weighed at all' (468). When I need an hour of Zen-like tranquillity, I now take refuge in one of the glass reading rooms on the ground floor of the Pratt Library. From there I can look through windows opening, not on the foam of Keats's perilous seas forlorn, but on the falling waters and pool of the Pearson peace garden. Contemplation differs so radically from research and practical learning that it is always an error, I think, to measure their achievements on the same scale. Julius Getman properly distinguishes academic publication, or the mere recycling of known facts, from two other activities: the scholar's contribution to the free play of ideas circulating in a culture, and the research scientist's discovery of new physical laws.

The contemplative withdrawal of the scholar from the modern world is well symbolized by the University of Toronto's Massey College, with its quasi-monastic interior and contemporary façade. Such oases of reflection should exist, not as elitist conclaves, but as small secluded Utopias for directing the independent studies of gifted students or for the leisurely pursuit of a scholar's own intellectual obsessions. Unfortunately, the centres of higher learning in a country like Canada, which loves to level its tallest trees and ensure a rough equality, are unlikely to join the supreme circle of great universities. Where can we find in our midst such enlightened conclaves as Harvard's Society of Junior Fellows or Princeton's Institute for Advanced Studies? When the most energetic exchange of ideas takes place in the waiting rooms of international airports by experts who are seldom to be found on campus, the great university at which Northrop Frye once talked in a local pub every Monday afternoon to his graduate student Peter Fisher about religion, literature, and Eastern philosophy has already faded into a distant memory and dream.

A picture of the kind Massey College provides is never exactly a portrait, but a utopian vision of what college might be like for a few qualified scholars. It recalls Henry James's idealized report of Oriel College in *The Galaxy*. Evoking a quiet dinner that he shared with some Oriel College dons in their common room, James praises their 'superior talk upon current topics, and over all the peculiar air of Oxford – the air of liberty to care for things of the mind assured and secured by machinery which is in itself a satisfaction to sense' (Novick, 369). As one commentator says, James's reminiscence of Oxford is also 'an image of loss,' of what had been denied to him (as it had been denied a century earlier to Samuel Johnson after a brief interval of study). James's portrait is 'like the one Virginia Woolf would paint fifty years later, almost on the same spot: a vision of a room of his own, of freedom from care' (Novick, 369).

Unlike journalism and computer engineering, scholarship and scientific research are both elitist enterprises. The figural interpretation of Scripture and the exploration of quantum mechanics are too arcane for most laymen to understand. But whereas scholarship is in principle as esoteric as a prophet who speaks in enigmas and sees through a glass darkly, the secrets of science are open to everyone who can conduct its experiments and understand its theories. Indeed cooperative group activity is to scientific research what solitude is to contemplative scholarship: its life-blood and soul.

If this distinction is valid, then grant forms that ask scholars to talk about their research partners and their teams betray a deep misunderstanding of what most scholars are doing. Too often a scholar is asked to be incompatible things. To the cunning and pertinacity of a grant-winner and competitor, he is expected to add the detachment of a god contemplating the world dispassionately from Mount Olympus. Combining the team activity of a scientist with the wisdom of Socrates, the modern scholar is asked to be a conniving Dr Grantly, a disinterested Arabin, and a contemplative Reverend Harding, all in one.

To Getman's three categories of research, scholarship, and academic publication, one might add a fourth category of practical learning, which includes the professional discourse of lawyers' briefs and court judgments. But as Abraham Flexner argues, successful lawyers make only minor contributions to the world's thought (276). Since a barrister's performance in a courtroom, however ingenious or mentally agile, seldom produces a masterpiece like Pericles' funeral oration, a parliamentary address by Edmund Burke, or Lincoln's Gettysburg address, it would be a mistake to confuse practical learning with either contemplative scholarship or pure research in the sciences.

A quantitative demand for publication may be appropriate in research institutes, where new discoveries have to be shared with other biochemists if stem cell research is to advance. But writing about Browning's or Tennyson's language is a contemplative activity. Scholars who study Hopkins's poems or Donne's sermons, who read them by day and meditate by night, are doing so because they love to, not because they are satisfying criteria transferred to the art of contemplation from a social or physical science.

Debates between scholars like David Bromwich and David Damrosch, who defend the conflicting claims of the individual and the group, are as irresolvable as the tensions between solitude and community that trouble Thomas Merton in his Kentucky monastery. Perhaps the solitude of the scholar-critic, alone with his books in a library, is the closest a modern Socrates can come to freedom from group conformity. But as Margery Sabin shrewdly observes, collaborative activity need not be collective. Projects that require a 'mastery of property law, of the economics of Victorian publishing, or of Lacanian psychoanalysis' may have to be subordinated to more purely verbal concerns that allow literary scholars and critics, even when collaborating with historians or philosophers, to refine the models unique to their discipline. An energetic defence of the 'intellectual, the aesthetic, and (not least) the professional identity of literary study' (100–1) may require scholars or critics to cultivate once again more 'supple and subtle reading habits' in themselves and their students (98).

A scholarship of integration may require broad reading across genres and even some collaboration of literary scholars with art historians and philosophers of science. But as David Bromwich warns, one should always beware of 'loss of confidence in areas where some real competence might still be fostered' (128). The new sophists are group narcissists, anti-traditional but authoritarian. They can never assimilate the great sustaining paradox of contemplative scholarship, the fact that it is 'a kind of solitude and a kind of company' in one (Bromwich, 130). Since solitary reflection and communication are both necessary, Bromwich argues that the scholar's contemplative discourse is best conceived as a 'conversation.' Defending the analogy favoured by Michael Oakeshott, R.G. Collingwood, and Richard Rorty, Bromwich observes that 'each speaker frames a statement in the belief that someone listening is capable of a reply.' Moreover, all 'the speakers have an interest in common that none can easily describe. And ... the good of conversation is not truth, or right, or anything else that may come out at the end of it, but the activity itself in constant relation to life' (132).

I have to concede, however, that the conversations one most wants to have seldom take place. I have tried for four decades to conduct a one-sided conversation with Robert Browning, one of the immortal dead. But the more I study him, the less I understand what he says. When I play with the idea of building long-range transmission devices that might allow my contemporaries to talk to posterity, I am chastened by the thought that nearly all scholarship or criticism is written to be superseded. Like conversations in common rooms, discussions at conferences designed to draw the most participants use platforms so level that for every scholar who engages us we are bored by a dozen pedants. Emerson says we need some gift of transcending time, as we transcend space, by choosing our partners in conversation from a wider brotherhood. But if a conversation with Plato, Socrates, Aristotle, or Plotinus were ever to take place, I fear it would resemble 'a definitive response to a performance of *King Lear*.' Such an experience 'would blow our minds' and 'effect an unimaginable transformation in our whole sense of reality' (Frye, 1988, 151). The integration in conversation that is genuinely desirable is generally *within* a discipline or *between* two closely related disciplines like literature and philosophy. Any communication between 'Romance philologists and solid-state physicists,' as Frye says, is likely to be a very 'unsubstantial Eucharist' (1988, 60). 'Intellectually,' he says, 'the world is specialized and pluralistic, and learning, like the amoeba, can reproduce only by subdividing' (1988, 152). .

Only rarely can scholarship produce the 'colossal simplifying vision' (Frye, 1988, 61) of a true prophet or revolutionist. When such vision occurs, it is usually when disciplines like geology and physics regroup, or when, as Frye explains, Freud emerges 'from a point of mutation at which psychology begins to turn into something unrecognizable to its scholarly establishment' (1988, 61). In combining vision and commitment, the best contemplative scholars are willing to take risks. Good students often make bad scholars, because they would rather transmit other scholars' ideas than devise arguments and theories of their own. Sometimes the neglected scholar can gain recognition by joining a scholarly movement of new historicists or feminists. But it is hard to strike a healthy balance between the often conflicting claims of the individual and the group. As Julius Getman warns, 'joining a scholarly movement reduces independence as to subject matter, conclusion, and even language – thereby reducing the likelihood of original research' (51). It is difficult for a group thinker or even a loyal disciple – a devout believer in Women's

Studies, a Marxist, or a behavioural psychologist – to be one of Getman's 'genuine intellectual chance-taker[s]' (120).

Perhaps the most gifted scholars are alienated from students and colleagues precisely because they recognize more clearly than anyone else that, in Julius Getman's words, 'professional recognition and status are different from and often in conflict with meaningful achievement either as a scholar or as a teacher.' Though 'membership on a distinguished faculty brings the former, it may actually discourage the latter' (262). Instead of writing another treatise for a learned professional group, Joseph Butler and Charles Darwin reach across traditional boundaries to produce such influential masterpieces as *The Analogy of Religion* and *The Origin of Species*. In communicating what one commentator calls 'devastating new ideas outside their own coterie' (Vidler, 95), groundbreaking scholars share with colleagues their boldest surmises and deepest misgivings or doubts. Instead of endorsing some predictable party line, they take their colleagues on adventurous expeditions of the mind.

Though Socrates uses his dialogues to demolish false knowledge, once he moves toward a true contemplation of friendship, piety, or virtue, he often replaces dialogue with an oracular monologue or myth. Despite many clichés about the magic properties of dialogue, the scholar *qua* scholar meditates in solitude on an often esoteric subject, which he ponders alone or perhaps in the company of a few close specialists and friends. For most deans and professors of public administration the organizational model for universities will always be the urban, industrial culture of modern business. But when the world is too much with the scholar, the university should also provide him with the equivalent of J.H. Newman's monastic experiment at Littlemore, a place where reflective souls can read, meditate, and even live together, alone and in silence.

In Frye's terse words, contemplative scholarship 'is esoteric, almost conspiratorial, and the principles of academic freedom require that it should be left that way' (1988, 60). One thinks of Newman, dismissed from his duties at Oriel College for trying to reform the tutorial system, emerging from his religious and educational experiments, not as an acknowledged leader or establishment figure, but solitary and alone, the quintessential outsider. Subsisting like Arnold's scholar-gipsy on the margins of Oxford society, Newman, according to his latest biographer, is 'the first great, and perhaps the most enduring, Victorian skeptic' (Turner, 641). Memories of Marshall McLuhan, brooding alone in a corner of the Gothic chapter house where the University of Toronto's graduate faculty met, or of Frye himself communing only with Blake or the prophets at the Victoria College high table, reinforce the impression that great scholars and humanists are often isolated in their own communities. Like Milton, Blake, or even Newman, they belong to sects or parties of one.

6

Practical Knowledge: Prometheus to Faust

Whereas Aeschylus thinks Prometheus deserves to be punished for stealing fire from heaven, Shelley celebrates him as a humanitarian hero. The medical doctor who uses biochemistry to cure disease or the engineer who uses technology to build higher condominiums and faster freeways is our new Prometheus. But in the triumphs of genetic engineering, the amassing of huge nuclear arsenals, and the mindless spread of urban communities, we are reminded that a Promethean overreacher like Faust has a sinister shadow side to his character. Ancient Greek and biblical texts from *Prometheus Bound* to Genesis warn against the forbidden knowledge of would-be gods like Faust, the black magician who bypasses time-consuming inquiry and research in exchange for instant knowledge and useful short-term results. In their attempts to reduce humane scholarship to a technique, experts in brain studies and behavioural psychology seem to be repeating Faust's mistake. Their description of human consciousness from 'the outside in' ignores the first axiom of all humane learning: that the meaning of a mind or soul is accessible only from 'the inside out.' In a reductive parody of the Socratic humanist and Prometheus's theft of fire from heaven, the new Prometheus produces the Frankenstein of human cloning and the many monsters of robotic technology.

I use the phrase 'practical discourse' to describe two distinct activities. Of primary concern are the attorneys and physicians who are said to practise law or medicine, in contrast to the philosophers who are said to profess logic or metaphysics. A discipline is practical if it encourages Prometheus to steal fire from heaven in order to find cures for cancer or build safer bridges. Of more general secondary concern is the practical quality of any discourse that makes meaning or truth a function of a word's consequence for action. Robert Frost's farmer is being practical when he defines 'home' as 'the place where, when you have to go there, / They have to take you in' ('The Death of the Hired Man,' ll. 118–19). There is also a large practical component in the discourse of analytic philosophy, which identifies meaning with function and distinguishes the performative use of words from their normal descriptive function.

A master of contemplative scholarship receives and transmits a revealed model, which is the secular equivalent of Moses' Law or a word made flesh. But in practical education, the deft use of an X-ray or cardiogram *demonstrates* or *shows* something. Like the skilful sketch of an artist's model, its enactments are often mute. A master of practical education is not exactly a showman, in the sense of being an actor or performer. Yet his 'teaching is ostensible.' As George Steiner says, 'it shows. This "ostentation," so intriguing to Wittgenstein, is embedded in etymology: Latin *dicere* "to show" and, only later, "to show by saying."' For a practical genius, 'only the actual life of the Master has demonstrative proof.' In Steiner's words, 'Socrates and saints teach by existing' (4). Unfortunately, most academies are so committed to the authority of a written model, which is their version of a sacred template, like the Tables of Moses, that they are deeply suspicious of any genius who refuses to commit his teachings to writing. Few academies would give Socrates or Jesus tenure.

One of the shrewdest defenders of practical discourse and knowledge is Michael Oakeshott, who opposes a life of practice and habit to the politics of rationalism.[1] Application and practice are a necessary antidote to the abuses of theory and abstraction. Concreteness is the strength of practical education, which associates 'thought with foresight and foresight with achievement.' Practical education 'gives theory, and a shrewd insight as to where theory fails' (Whitehead, 1949, 64). The utility of practice in medicine, which applies the scholarship of scientific discovery to human disease, becomes a scourge only when society prevents the university from making utility, not the natural by-product of scholarship and research that it ought to be, but an end in itself. Another abuse occurs when a model of discovery that is appropriate to the practice of normal science is illegitimately imposed on the practice of history or philosophy, for example, where different models apply.

As Oakeshott observes, practical knowledge 'exists only in use.' It 'is not reflective and (unlike technique) cannot be formulated in rules' (1991, 12). Technical knowledge about an English sonnet's voltas, rhyme scheme, and concluding couplet is easy to learn. But knowledge about how to write 'Shall I compare thee to a summer's day?' or how to produce discerning criticism of the sonnet is practical. Oakeshott believes that if such knowledge can be imparted at all, it can be imparted only by example. For it 'exists only in practice, and the one way to acquire it is by apprenticeship to a master' (1991, 15).

A scholar or scientist who is well educated (as opposed to well trained) will possess the practical knowledge that tells him when his models or paradigms are 'leading him astray' and how 'to distinguish the profitable from the unprofitable directions to explore' (Oakeshott, 1991, 15). Though rationalists protest that 'practical knowledge is not knowledge at all,' it has the same value for Oakeshott that poetry has for Wordsworth: it represents the 'breath and finer spirit of all knowledge.' As the leaven of mere technique, practical knowledge is personal and concrete, opposed to the tyranny of any abstract doctrine or formal set of rules.

Presumably all competent scholars and scientists have a practical working knowledge of the models in their disciplines. Whereas a sophist studies a subject in order to contradict and refute an opponent, a wise Socrates knows how to weigh, consider, and use what he knows. 'Crafty men contemn studies,' Bacon says, 'simple men admire them, and wise men use them.' Since studies seldom 'teach their own use,' all genuine learning includes a practical component, which Bacon praises as a 'wisdom' outside studies 'and above them, won by observation' (128–9). Since no discipline, however, can ever adequately impart this practical knowledge in a set of directions, maps, or rules, all that the textbooks or the survey courses can teach is the mere ghost of a model, or what Oakeshott calls a 'technique.' Every such handbook or set of rules has the status of Machiavelli's *The Prince*, which offers nothing but a 'crib to politics.' In default of a genuine political education it provides mere 'training,' or 'a technique for the ruler who has no tradition' (Oakeshott, 1991, 30). Deploring a recent obsession with rules and theory, Oakeshott advocates a return to a less self-conscious culture of habit and custom.

In a scientific age, it is a dangerous mistake to assimilate all learning to the research model. All research is a form of scholarship – the scholarship of discovery, if you will – but not all scholarship is research. The scholarship of a philosopher or literary critic is likely to be contemplative. And as Judith Shklar reminds us, some important contributions to knowledge are examples of the scholarship of teaching. At the University of Toronto the research fallacy is perpetuated in the very name assigned to the office that handles applications for academic grants: the Office of Research Administration. Even the name of Canada's Social Sciences and Humanities Research Council repeats this error. For centuries medieval scholasticism subordinated the research model to the scholarly model. But today the opposite tyranny of assimilating scholarship to research is the greater menace. Too often the research model of the natural sciences, including its quantitative methods of evaluation and its preference for team projects over individual investigations, is imposed on branches of humane study and scholarship where it is uncongenial or even deadly. Should a reasonable humanist adapt himself to the ascendant research models? Or should he be unreasonable and persist in trying to adapt the world to himself? It seems to me that progress often depends on unreasonable scholars.

Most explanatory discourse combines elements of practice, discovery, and contemplation which, if left unmodified, might produce mere vocational education at the practical end of the spectrum and purely monastic education at the contemplative end. The play impulse is found in its purest form in aesthetic criticism of the arts; when combined with the science of genres or with the power plays of a 'political unconscious,' aesthetic criticism incorporates the competitive impulse of a game. Just as a practical occupation like business administration seeks status as well as knowledge by aligning itself with a centre of higher learning, so a community of faith committed to a religion of the book or to a specific liturgical practice may seek academic prestige or doctrinal stability by aligning itself with a school of theology. The more practical the school, the more it resembles a seminary: the more liberal, the more it resembles a college or university.

Sometimes the best way to be practical is to be impractical. Often the moments when we see into the life of things or die for a few instants to the rush of time take us by surprise when we are relaxing on a hike or taking a coffee break. Perhaps, as Wallace Stevens muses, the 'truth depends on a walk around a lake' ('Notes toward a Supreme Fiction,' 'It Must Be Abstract,' 8.3). An unconsidered or unrehearsed life may expose a scholar to the shock of an unexpected encounter that transforms his understanding and changes him from within. Oliver Wendell Holmes tried to practise law as if he were writing a poem: perhaps one should practise a profession as if one were taking part in a stage play or a chess game. Since a designer with a sense of style knows how to economize his materials, Whitehead believes it is just as important for an engineer or an architect to cultivate an aesthetic sensibility as it is for an artist. And just as a profession like engineering or law will try to cultivate a speculative or theoretical side by associating itself with a faculty of arts and science, so even a contemplative or speculative discipline like theoretical physics will also reverse this process by testing its theories in a laboratory.

A New Critic may test a favoured doctrine about the ironic content of great poetry by passing it through the crucible of a lyric by Wordsworth or an ode by Shelley that seems at first to disprove the doctrine. Or a philosopher may conduct a thought experiment about the sound of a falling tree in an earless forest. Both scientists and scholars use, discover, and invent their paradigms. And yet there are important differences between their practices and styles. A physicist is less interested in the experiments Newton conducted in the eighteenth century to establish the laws of motion than in how contemporary scientists use or modify his discoveries. The scholar, by contrast, is a collector. He is not just a user and discoverer of new models, but a custodian of past cultures who proves the truth of Rilke's saying that rightly to commemorate what has died is also to magnify what lives. Dead paradigms acquire posthumous life in the scholar's histories of science or in his chronicles of what C.S. Lewis calls a 'discarded image.'

Practical discourse tends to equate meaning with use: the truth of a proposition is to be found in its consequence for action, as pragmatists have always argued. The most spirited attacks on impractical reflection and leisure come from the pragmatists William James and John Dewey, who share Karl Marx's impatience with philosophers who 'have only interpreted the world' instead of changing it (quoted by Pelikan, 157). Marxism, however, is a secular but essentially religious version of the classical religious dream. And its commitment to change should never be achieved through proselytizing. Since the university is not an office of propaganda, it cannot be a party headquarters or a seminary. Disinterested research seldom has as clear and specified an outcome as the philosophic inquiries of a scholastic theologian or as Marx's investigations into the history and economics of the British coal industry. As Pelikan says, 'it is cheating to play the scholarly game with marked cards even if one can win that way; complete objectivity may be impossible, but scholarly honesty is not' (162).

In professional education, Donald Schön's 'reflective practitioner' must often build a bridge from the terra firma of 'rigorous professional knowledge, based on technical rationality,' over the 'indeterminate, swampy zones of practice that lie beyond its canons' (Schön, 1990, 3).

Schön gives two examples of the need to apply the well-formed structures of professional education to 'real-world practice.' In the first case, a physician identifies a cluster of symptoms that she cannot associate with a known disease. In the second case, 'a mechanical engineer encounters a structure for which he cannot, with the tools at his disposal, make a determinate analysis' (1990, 5). In both instances, since the case is unique and not 'in the book,' the professional must experiment and improvise. At other times conflicts in professional education arise from conflicting values. An efficient industrial process may deplete the ozone layer. An expensive medical technology like kidney dialysis may stretch a nation's ability to invest in other forms of medical care. As Schön concludes, such 'indeterminate zones of practice – uncertainty, uniqueness, and value conflict – escape the canons of technical rationality' (1990, 6).

In architectural education, as in some branches of scholarship and science, professors embody in their own reflective practices secrets of artistry or technique they cannot clearly formulate or share. As Schön observes, professors of architectural design find it difficult to tell their students 'what designing is, because they have a limited ability to say what they know, because some essential features of designing escape clearly statable rules, and because much of what they *can* say is graspable by a student only as he begins to design' (1990, 100). Using a shorthand that may seem 'obscure or complex' to anyone who is not trying to design an office tower or a museum, Schön's instructor has more in common with a mentor or a coach than with a professor in secure possession of a theory to expound.

To learn the arts of architecture, medical diagnosis, or psychoanalytic practice is to master what Yeats calls a 'secret discipline.' Though Schön concedes that 'we tend to think that artists create things and practitioners of other professions deal with things as they are' (1990, 217), he prefers to regard architectural designers, psychoanalysts, and lawyers as 'worldmakers' who use the tools of their profession 'to impose their images on situations of their practice.' In Schön's view, 'a professional practitioner' is, 'like the artist, a maker of things' (1990, 218). And the artistry of his practice, whether it be in architectural design, musical performance, or psychoanalysis, is better taught by a coach's tips than an instructor's lectures.

Schön's reference to the artist as 'a maker of things' recalls Aristotle's contrast between doing and making. The dictum that we cannot teach what is most worth learning – how to live a good life or 'do' philosophy – is a legacy of Aristotle's contrast. The embodied wisdom of Socrates, Einstein's way of 'doing' physics, and Shakespeare's extraordinary life, which Keats calls a work of continual allegory, are forms of doing or performance in Aristotle's sense. But like Aristotle's intellectual and moral virtues, such performances are too lofty and godlike to be taught. At the opposite end of the scale are forms of making like carpentry and cooking, skills which, though clearly teachable, are too banausic to include in the academy. Most practical knowledge lies somewhere between the two extremes of Einstein and Mr Fixit, Shakespeare and the French chef. A practical science like engineering can teach a student how to build a bridge. And to the degree that architecture is a science, it can tell a student how to design a museum. But architectural design is also an art, and to be an artist is to act and behave like one. The architect's ability to *do* as well as *make* involves not only a production of blueprints but also a more mysterious skill, a secret discipline, that Schön never quite manages to explain.

Curious as it may seem, a complex but more lucid version of Schön's difficult idea of 'reflection-in-action' can be found in John Grote's idealist theory of 'knowledge of acquaintance.' In an idiosyncratic but original contribution to nineteenth-century epistemology, *Exploratio Philosophica*, Grote argues that to grow and prosper an energetic mind must draw sustenance from two sources. It must accumulate sensory experience (or 'knowledge of acquaintance'). And it must abstract from this experience a formal set of rules, called 'knowledge of judgment,' which it can then apply to new experiences. Genuine learning in science or scholarship cannot be the mere 'knowledge of acquaintance' which John Grote describes as the 'rubbing of ourselves against fact' (148). Nor can it be only that knowledge of judgment which, in abstracting rules from experience, is too often a mere bloodless paradigm or ghost. As Grote insists, learning is a result of 'these two processes in conjunction' (148).

Browning's Pope says that theology is continually trying to 'Correct the portrait by the living face, / Man's God, by God's God in the mind of man' (*The Ring and the Book*, 10.1873–4). Such a renewed knowledge of acquaintance is necessary if the theological language of judgment is to remain in touch with human experience. Or as Grote says, 'thought and fact may be said roughly to correct each other' (148). He concedes, however, that the two views of knowledge cannot readily be combined. For if knowledge of acquaintance is 'a mirror reflecting the universe,' knowledge of judgment is 'an eye reflecting [it]. The first is true, but it is not knowledge; the second is knowledge,' but until it has been corrected and refined, it does not merit the prize word 'truth' (149).

Far from being an oxymoron or contradiction, the artistry of a professional practitioner is the only knowledge that a medical doctor, a lawyer, or an architect can hope to possess. Like Schön's 'reflection-in-action,' Grote's 'knowledge of acquaintance' consists of informed practices and habits rather than of mere precepts or theories. Such knowledge brings a strong increment of personal aptitude and talent to a collection of otherwise dead or inert facts. Closer to the wisdom of the contemplative scholar than to the factual information of the technician, 'knowledge of acquaintance' involves an exercise of those non-teachable habits that make up the daily artistry and science of Schön's psychoanalyst and architect. Not confined to professional education, 'knowledge of acquaintance' is common to the medical doctor who correctly interprets the symptoms of a disease, to the attorney who uses legal precedents to defend a client, and even to the literary scholar who recognizes in an epic simile by Tennyson an echo of a simile in Milton, which itself echoes a still earlier one in Virgil. We come to know a poet or a genre by moving with ease among many works. Such 'knowledge of acquaintance' is not progressive or linear but what Frost calls a form of 'circulation.'

All Socratic wisdom is based upon such cycles of reflective practice and habit. But when habit petrifies, it turns into a mere parody of wisdom. Socrates attacks in sophistry the second kind of cycling, with its mindless rotation of dead habits and prejudices. Professional practice is comparably blind only when drug companies, political lobbies, or 'evangelists' of commerce 'with a patron in their pocket,' as Oakeshott says, use threats or bribes as agents of repression. True custom or habit 'is always adaptable,' he notes, 'and susceptible to the *nuance* of the situation' (1991, 471).

The lawful marriage of practice and theory is under greatest stress in the medical and law schools. In theory, a medical student should treat patients as people. But in practice the student must reverse this priority by learning to see everyone as a patient. One Harvard medical student, Ellen Lerner Rothman, describes how she and her classmates were first taught to humanize medicine. But then in second year, without losing their compassion for patients, they were trained to look 'through medical eyes, no longer clouded by oblivion' (106). Though in some sense satisfying, Rothman admits it was also oddly disturbing when 'all the illnesses I studied in school suddenly seemed to materialize in the people around me, who I had always before assumed were healthy' (105). Gradually, the goal of patient-doctor interviews began to shift from reassuring the patients to a gathering of concise medical histories.

In trying to adjudicate the claims of practice and theory it is always difficult to determine how far the pure biochemical research of the medical scientist or the contemplative discourse of the philosopher of law can be tolerated in the professions. It is not axiomatic that the disinterested pursuit of cell biology or legal theory has any justifiable claim on the first fumbling attempts of a medical or law student to master the rudiments of her profession. Too often, as Derek Bok complains, the student's youthful enthusiasm 'is battered by the grinding effort to mount toward the first step of professional competence' (1986, 74). The problem is made worse by the competing claims of the scientist and the clinician or the legal theorist and the advocate. Several partners are often competing for allegiance: the academy, the students, the profession itself, and the larger society to be served.

Professors of law and medicine often reserve their greatest admiration for colleagues who devote their lives to academic research. These scientists and scholars are paid to conduct experiments in laboratories or pursue inquiries in archives, where they can seek cures for cancer or devise new approaches to labour law that demonstrate exceptional mental agility and powers of endurance. But practitioners of medicine and law may also admire more practical skills. They may place a premium on physical dexterity in brain and heart surgery, on psychological acumen in psychiatry, or on other talents like patience and fortitude that are less intellectual than moral. Students may also demand training in ethics and on the legal implications of performing intricate eye surgery when the chance of total success is problematic. The practice of a profession may be further complicated by such external social forces as the threat of malpractice suits, which hang like a sword of Damocles over many operating tables, and by the financial constraints imposed by governments, insurance companies, and HMOs. The society to be served by lawyers, doctors, or business executives may also extend or withhold its financial support in order to control the supply of graduates entering the profession.

In medical schools the same professor may be asked both to teach and do research, to train doctors clinically while conducting experiments in biochemistry and cell biology. It might be more efficient to hire two doctors, one to promote the research and the other to apply the discoveries of research to patient care. Even when both a medical scientist and a clinician can be hired, however, a dean of medicine may have to heal a rift 'separating' the 'faculty into two distinct cultures, which often clash with each other over money and space' (Bok, 1986, 87). Despite the risks of diffusing the students' energies and diluting their knowledge, the rewards of successful integration in a professional school like medicine, which is 80 per cent interdisciplinary, are also proportionately great.

But if medical and law students have to absorb vast volumes of information, is their education truly liberal? Does it train the mind or develop mere physical stamina? Perhaps, as Derek Bok suggests, it is more 'akin to memorizing the entire Manhattan telephone book – backward.' Bok observes that 'great chess masters are people who keep in their heads an enormous number of prior games, which they can use to decide what to do in any situation they encounter. Conceivably, doctors [as well as good literary critics, historians, philosophers, and teachers] work much the same way' (1986, 91). One hopes that no ultimate conflict exists between the art of the chess master and the moments of beauty that come to the brain or heart surgeon who reaches the pinnacle of his profession. To practise law, medicine, or engineering in a liberal spirit is to take delight in the expenditure of energy and talent for their own sake. A surgeon's great triumphs are seldom witnessed by anyone outside the operating theatre. But anything from fishing to surgery that is perfectly executed is beautiful. Grace comes by art, and art comes by discipline which is never easy and may take years to acquire.

The terrible news of a friend's death makes an impact. But news that I have terminal cancer, provided it is accepted and believed, induces a tremor. How should a medical doctor tell me I must die – and die soon? There are no rules for acquiring skills in psychological counselling or in breaking bad news. Crafts can be taught, but habits of continuous learning and apprenticeship contain large intuitive components.

A certain mystique surrounds the whole controversial question of how much students learn in professional schools. It is easy to boast about the aptitude scores of an entering class of medical or law students but difficult to say just how 'much these students actually learn after they arrive' (Bok, 1986, 3). Even grade-scores and other admission criteria are 'only moderately useful in predicting a student's record in professional school' or in later medical or legal practice (Bok, 1986, 122). Oddly enough, the leisure to refine one's art and enlarge one's thinking may be best utilized by successful doctors and lawyers who return to medical or law school in their mid-forties to absorb new knowledge. Such a seasoned professional is more likely to make important discoveries in the next twenty years than a student who is only eighteen or twenty-two.

But the ultimate faith of all professors, who are paid 'to enjoy a life of learning' (Bok, 1986, 6), is that, even in a medical, law, or engineering school, knowledge is always important as an end in itself and its mastery is always valuable. Professors like to think that a professional education, not less than a liberal one, promotes the qualities of mind and temperament associated with Socratic humanism. These simultaneously tough but fragile virtues include free inquiry, intellectual honesty, trust in the rationality of the universe (even if it includes large pockets of mystery), and a readiness to communicate the results of research – if only as a longstanding invitation to be corrected or proved wrong.

Though Socrates was a great teacher, he lived or embodied his truths instead of publishing them. He felt that any theories he might extract from his dialogues would be valueless out of context. Socrates therefore seems to validate Newman's troubling claim that 'to discover' theories and 'to teach' habits of mind or intellectual practices 'are [two] distinct functions' (quoted by Pelikan, 91). But if Socrates is the patron saint of professional education, whose embodiment of ideas subordinates theory to practice, his advocacy of an examined life serves as an equally stern rebuke to any professional schools that would sacrifice practice to theory. At some teaching hospitals patients with rare diseases are valuable only because, like exotic animals in a zoo, they provide case histories and data for highly specialized research.

The more we try to determine the standards of good professional education, the more complex and vexing they become. Though medical research usually feeds off the bounty of scientific academies and generous funding agencies, it should be remembered that Frederick Banting discovered insulin in an ill-equipped laboratory. Odd as it may sound, adversity itself may be the best teacher. Some research projects are the slow and painful product of solitary labour in a sparsely planted vineyard. Austerity may whet the mind of a medical scientist or make the collective will of an underfunded research team keener.

We must also beware of confusing the size of a professional school with its quality. Nor should we be seduced by an alluring stock-market method of comparing them. Each year Canada's *Maclean's* magazine ranks the country's professional schools and colleges in its so-called 'poll of excellence.' But this annual ritual, usually scorned by universities that score low in the poll and embraced by those that do well, is a futile exercise based on three fallacies. It wrongly assumes that all pro-

fessional schools and colleges *can* be excellent; that they would *want* to
be 'excellent' even if they *could*; and that excellence has a common
meaning for all constituencies. Even when its faculty, students, and capital resources are mediocre, each university claims to be excellent in a
different way. It is difficult to see how a shared excellence can be unique
(or one of a kind) and yet still exhibited as a property common to every
professional faculty and school.

A university that laboriously plans for excellence every five or six years
may be too diverted by navel gazing to exhibit excellence or even recognize it when, like a thief in the night, it pays a rare and furtive visit. To
plan for greatness is the subtlest of temptations. For like Becket's desire
to be a saint in *Murder in the Cathedral*, it is seldom compatible with the
exalted condition it seeks. What distinguishes, not merely a community
of Trappist monks, but profundity itself from platitude is a silent commitment to values that are lived or embodied rather than blared forth
on a trumpet.

Prometheus and Faust are not movie stars. And the city of art and culture is not Hollywood. By becoming an academic equivalent of the
Oscars, even prizes awarded by royal societies and funding agencies
deform higher learning. Like most media attention, a bid for publicity is
always a pall. It smothers the fire that is struck from the flint of a hard-
resisting discipline or received as a gift from the Muse. As John Metcalf
says, 'rarely do we find people taking the long view, patiently comparing
books from the present with books from the past. Rarely do we hear reference to the 'crafte so longe to lerne.' The universities, once a countervailing influence, seem to have abandoned any public role' (281).
Indeed most protocols of excellence appeal, not to the authority of a
learned tradition, but to the commercial analogy of a spiralling stock
market in an ever-expanding economy, where bigger is better. Unfortunately, this analogy mistakenly assumes that all professional achievers
want to imitate the giddy rise of Prometheus, who is always trying to surpass himself. Though a stock market investor wants this year's profits to
exceed last year's, it may sometimes be preferable to achieve stable equi-
librium. Who wants to be CEO at Enron, prince of the morning one day
and Lucifer the next, hurled headlong through space to a bottomless
perdition of insolvency and ruin? It is better to be the wizard Gandalf
than a fallen Prometheus like Sauron, dark Lord of Mordor. But a
scholar would rather be a hobbit in the shire than either.

Since professional education in business and law schools operates on
trust, it is vulnerable, like capitalism itself, to dishonest inflation in

everything from grades to self-promotion. I doubt if either professional schools or popular capitalism can survive for long without moral regulation. Like stock value and justice, the word 'excellence' is slippery, because it is charged with strong emotive meaning even when any intelligible descriptive meaning is hard to discern. 'Excellence' tends to have one meaning for the law professors and medical doctors who are teaching; another meaning for the students and interns who are learning from them; and still a third meaning for the deans of law and medical schools, who are calculating the market value of the two other forms of excellence in order to justify a hike in the tuition fees.

Competition may be the god of every market economy. But professional education combines too many intellectual and practical skills to be judged like a stock option, a music festival, or even a figure-skating championship. Nor does it resemble such commercial products as a condominium or a car, whose buyer can accurately assess its economic value. We must concede an irreducible minimum of ignorance in this area. Medical doctors and lawyers may feel gratitude to schools that admit and license them. But as Bok asks, do consumers 'really know how well various universities are doing in achieving the educational results they consider most important? If not, what assurance do we have that competition by universities for the favor of these groups [or competition among students for admission to the more prestigious schools] will improve the quality of education? The more one ponders these questions, the more troublesome they seem' (1986, 34).

In trying to respond to the often conflicting demands of students, scientists, scholars, clinicians, practising lawyers, alumni, government agencies, and charitable foundations, North American professional schools may be pulled in too many directions at once. The competition among many contentious interest groups may have the desirable effect of preventing control by any single faction. It may even minimize the negative effect of such serious errors of judgment as a surplus or deficit of attorneys and family physicians. But competition may also be unprincipled and wasteful. To compete for the funding of psychiatry programs by large pharmaceutical companies, a medical school may withdraw a job offer to a professor who warns of the dangerous side-effects of Prozac. And medical scientists may have to spend 20 per cent of their time preparing grant applications instead of pursuing their research on cancer or AIDS. Even when competition appears to spur efforts in the search for a cancer cure, its beneficial effect on law professors is more doubtful. As Bok says, competition may make legal scholars or professors in the humanities 'too susceptible to intellectual fads or excessively inclined to take unorthodox positions in an effort to be "original"' (1986, 33).

It is also uncertain how well computer-assisted instruction can educate young lawyers and doctors. The best opportunities come from computers that conduct Socratic dialogues, that design 'expert systems,' or that can even simulate live patients. But there are limits to progress in each of these areas. In a genuine Socratic exchange, students can devise their own questions, not simply choose from a predetermined list. Unfortunately, 'expert systems' do not force students to solve problems for themselves. They simply ask the student to supply the data and then observe how the computer responds. Bok believes that computer simulations in medical schools work best when constructing hospital environments too dangerous, expensive, or inaccessible for students to experience at first hand (1986, 149).

Computer technology is an efficient servant but a bad master. A wise administrator may have to rescue a scholar from the unmanageable good temper of a dean who believes Utopia has been achieved when a word processor is installed in every office and every faculty member has created a web page on the Internet. Equally distracting are the students who clamour to communicate with their professors by e-mail, a poor substitute for conversation. Computers are most human when they send student records to Bangladesh or whimsically trigger a snow alert on an August afternoon. Like a repeating fraction launched on an infinite regress, computers are most endearing when they prove fallible like the rest of us. Whether through power failures or whiteouts, education should allow students their 'moment to reflect.' As Frost says, the delay – 'the time out' – confers value on the sloped head, the gesture of a mind composed for thought.

It may be expedient for business and medical schools to build their big temples and clinics close to the offices of political economists and to the laboratories of biochemists and biologists. But to seek knowledge in the shadow of a great medical school which evicts scholars from offices in its information centres, attracts many of the best graduates to its ranks, and swallows up the budgets of whole science and humanities departments can demoralize everyone who is not in medicine. One can see why Princeton has no medical school, and why Harvard and Cornell separate their medical faculties from scholars and scientists on the main campus. In the race for prosperity and improved technologies, it is tempting for professional schools to neglect the pursuit of knowledge for its own sake. Even at the risk of introducing conflict and tension, it is important that they do everything in their power to preserve exacting standards of self-criticism, service to the community, and mental rigour. It is also important that every faculty make room for the brightest people, who should not be excluded simply because they are socially inept, graceless, or eccentric to the point of doubtful sanity. No one would ever have called Marshall McLuhan or Northrop Frye companionable colleagues. They were solitary and sometimes touchy, but as hard to ignore in their implicit judgment of lesser minds as a prophet or a conscience.

Derek Bok asks why university professors who 'may well have won a Nobel prize (or its equivalent) and made great contributions to society, ... should earn so much less than worthy professionals in other fields' (1993, 156). It seems unjust that a successful football coach should be paid 'a sum several times greater than the salary of a senior professor' (1993, 157). It is at least incongruous that an 'assistant professor of neurosurgery at Cornell' should earn 'more than nine times the salary of his own university president, who is widely regarded as one of the nation's most accomplished academic leaders' (1993, 223). Bok takes heart from the fact that, regardless of what they earn, most of the small band of truly original thinkers who are capable of adding significantly to knowledge 'have strong intellectual commitments and will probably enter academic life in any case' (1993, 160).

A scientist who prefers commercial success to medical research, or a scholar who stays in the academy but whose finish is never worthy of his start, faces the bleak fate of George Eliot's Lydgate, who breaks the most important promises, the ones he makes to himself. Eliot's lighthearted remark that Lydgate wrote a treatise on gout, 'a disease which has a good deal of wealth on its side' (610), cannot disguise the bitterness of the doctor's own black humour. His wife is his basil plant, he jokes, a herb which 'had flourished wonderfully on a murdered man's brains.' Ironically, it is Lydgate who was implicated in the homicide of Raffles. But the true murderer is Rosamond, who destroyed the noblest ambition of her husband, his dream of becoming a medical pioneer, a Victorian Vesalius. Outwardly, Lydgate is a success, even to his wife, who values income above achievement, and who is provided, not with the 'threatened cage in Bride Street,' but with all the 'flowers and gilding, fit for the bird of paradise that she resembled' (610). As charming as a seductive snake, the metaphor focuses on similarities between marriage and a prison, made tolerable only by the gold and tropical flora that cater to the whims of a rare exotic bird. And yet coming in a coordinate clause after a tribute to Lydgate's well-remunerated medical skills, the judgment is crushing. Its disapproval of Lydgate is no less devastating for being understated and all but buried in the inventory of apparent successes that surround him: 'but he always regarded himself as a failure; he had not done what he once meant to do' (610).

As Bok eloquently reminds us, 'a long, almost unbroken tradition of secular and religious thought informs us that ... a preoccupation with material gain can produce [neither] a deeply satisfying existence [nor] a life we look back upon with pride' (1993, 297). And yet the growing prestige of professional education does, I fear, pose troubling threats to scholarship and science. Professional schools try to limit their numbers by rejecting more applicants than they admit because society already has more lawyers and medical specialists than it needs. Nevertheless, as more doctors and lawyers are admitted to high-paying professions, litigation tends to increase and doctors are tempted to prescribe more tests and recommend more surgery to keep them in business. Most disturbing is the increasing proportion of bright undergraduates who desert research, scholarship, and teaching for the more lucrative professions.

Centres of higher learning have to face the unpalatable truth that some forms of practical education cannot be enlisted into Mortimer Kadish's 'liberal service of the self' (124). However useful in themselves, such professions as hotel management and animal husbandry are expensive distractions from the proper function of a university education. With the rigour of Moses denouncing the worship of the golden calf, Abraham Flexner tries to banish all schools of business administration and journalism from every campus on the planet. The skills required by a business executive or journalist cannot be taught in a liberal enough spirit to be truly professional. 'Modern business,' Flexner concludes, 'is shrewd, energetic, and clever, rather than intellectual in character; it aims – and under our present social organization must aim – at its own advantage, rather than at noble purpose within itself' (164).

In my judgment, Flexner's strictures are excessive. In the first place, too empirical or self-serving a pursuit of learning is no argument against its proper use. Though training in fruit-growing or meat-carving is too practical to include in a professional curriculum, even medicine and law cease to be professional when they rank profit above discovery or expediency above principle. Even a traditional subject like physics or philosophy can turn the academy into a cafeteria offering only junk food for the mind unless the menu is attractively advertised, the contents imaginatively organized, and the food itself intelligently selected and consumed.

In the second place, Flexner seems to forget that a practical science can be just as worthy of serious study as philosophy or theoretical science. Though one does not have to be a Marxist to concede that capitalism can corrupt as well as energize a culture, what confers intellectual distinction on an academic discipline like economics or business administration is not its subject matter as such. Of more importance is the capacity of a discipline's professors and students to raise searching questions and probe enduring issues; to assemble the pertinent facts and information; to appraise the values involved; and to arrive finally at an informed judgment.

Nevertheless, Flexner's criticisms of vocational education contain a nucleus of truth. Even when universities filter applicants through a sieve of rigorous interviews and tests, they tend to limit higher learning to computational or verbal skills that lead directly to professional success. As T.S. Eliot says, 'the prospect of a society ruled and directed only by those who have passed certain examinations or satisfied tests devised by psychologists is not reassuring; while it might give scope to talents hitherto obscured, it would probably obscure others, and reduce to impotence some who should have rendered high service' (1948, 101). The 'nightmare of the purely instrumental mind' reaches a zenith of technological absurdity in Mortimer Kadish's picture of a university graduate 'educated to be a compact disk ... loaded down with highly useful programs' (13). When mechanical engineering or plumbing 'boasts that it alone has, through its discoveries, advanced human knowledge, it is as if,' in Schopenhauer's words, 'the mouth should boast it alone keeps the body alive.' The training provided by a vocational education 'is related to thinking as eating is to digestion and assimilation' (Schopenhauer, 92).

As early as 1848 Ralph Waldo Emerson deplores an American obsession with vocational training and utility. A premature and indecent attachment to professional education sets an American undergraduate apart from his English counterpart. The American student is less devoted to scholarship than to education that equips him for politics and trade. In England, by contrast, a public school 'excludes all that could fit a man for standing behind a counter' (Emerson, 1908a, 104).[2] In declining to teach modern languages, Oxford even refuses to help its sons 'fight the battle of life with the waiters in foreign hotels' (Arnold, 1932, 86). Emerson himself is too progressive to approve all the 'gales that blow out of antiquity' and direct 'the vanes on all [Oxford's] towers' (1908a, 106). Though he would like to see more Oxford professors filling up their vacant shelves as original authors, he admits that most universities, because of their fetish for tradition, are as necessarily hostile to genius as churches and monasteries are to wayward young saints. Emerson speaks with some disdain of the Oxford student who is a mere machine or 'studying-mill.' Compared to an American dilettante, he is as a 'steam-hammer' to a 'music-box' (1908a, 103). But though it is possible to pursue even a liberal education in an illiberal spirit, Emerson's admiration for an enclave of genuine scholars, enjoying the respect of all cultivated countries, should not be minimized. 'When it happens that a superior brain puts a rider on this admirable horse' of classical education, 'we obtain those masters of the world who combine the highest energy in affairs, with a supreme culture' (1908a, 103).

Flexner astutely observes that 'the world gets its stimuli from the genius, but it lives on talent' (352). Though he agrees with Emerson that the university can do nothing to foster genius, it has a responsibility to provide a home for talent. Instead of squandering vast sums of money on manicured lawns, huge stadiums for the alumni, big mansions for the president, and Olympic-size pools for the athletes, universities should attempt something more modest: they should build adequate library and research facilities for every apprentice scholar or scientist with a will to learn. Instead of competing with the football teams for star performers, and placing other players on waiver or short-term contracts, universities should provide enough compensation and security to make teaching and scholarship, in Flexner's words, 'a decent and possible profession for men of brains and taste in sufficient numbers' (361).

David Bromwich provides a devastating critique of efforts to make contemplative scholarship conform to professional models. '*All thought is scientific; professionalization alone suffices to turn art into a science*; therefore, for us, *professionalization makes thought scientific*' (191). Bromwich uses the hideous noun 'professionalization' to deride a pernicious group narcissism beloved of teacher training institutes, some faculty associations, and scholars like Gerald Graff, who offers the choice axiom 'Professionalization makes thought possible.' What 'makes thought possible' is not such 'bureaucratized narcissism,' Bromwich contends (195), but scholarly solitude and contemplation, the 'high commitment to intelligence,' and the habit of 'reserved engagement' inseparable from the pursuit of any worthwhile project. Students with no talent for science want to be medical doctors, and many failed scholars aspire to be journalists or politicians. The veneration of professional education as an ascendant god or idol is a cancer ready to eat out and destroy the soul of any university.

Unfortunately, Socrates is an easy genius to parody but difficult to imitate. And nothing can prevent the parasitic multiplication of professional sophists who thrive on make-believe scholarship and science. As Flexner says, 'the moment a real idea has been let loose, the moment technique has been developed, mediocrity is jubilant' (25). But Flexner is willing to tolerate sophists and professional pedants for the simple reason that the conditions that nourish the wise Socrates and productive scientist are precisely the conditions that favour somnolence. He concludes that 'it does not so much matter that some persons go to sleep, provided enough others are wide awake and fertile, at the maximum of their powers' (25). Pedants like George Eliot's Casaubon can lose their way in a maze of pen scratches as long as a few innovators like Lydgate survive. Since Socrates and the sophists are always in conflict, a university must be flexible enough to avert chaos without inhibiting creative thought.

Flexner believes that a medical faculty would be ruined if, instead of pursuing its own scientific research, it tried to live parasitically off the pure research of others. 'The professor of medicine is primarily a student of problems and a trainer of men. He has not the slightest obligation to look after as many sick people as he can' (16). The intellectual adventure of keeping abreast of the scientific literature and of making his own contributions to the control of West Nile virus or SARS takes precedence over any imaginary duty to treat numerous patients. The scandal of modern medicine is not only the failure of its great crusades against cancer and AIDS but also its profound ignorance of the biological cause of most disease. Instead of assigning rheumatoid arthritis or multiple sclerosis to such pseudo-causes as stress or faulty genes, medical scientists must devote time and energy to explaining why the joints of arthritic patients become painful or inflamed and why the nerves of sclerosis victims lose their protective sheaths. Medical science must understand that each pathology has its own bizarre symmetry: even a deadly SARS virus may be renegade DNA. Until medicine abandons its futile pursuit of isolated cures and thinks of disease as inverted order, keyed to the mad resolve of a few delinquent bacteria and cunning genes, it will remain in the Stone Age.

Albert Einstein says that 'we should take care not to make the intellect our god; it has ... powerful muscles, but no personality' (quoted in Charlton, 37). Any professional school of engineering or medicine will appear culturally depleted and thin if it slights the personal knowledge and commitments that remain the hallmark of any education that is not just technological but 'broadly and deeply scientific' (Flexner, 21). Professions derive their dignity and authority from the 'promotion of larger and nobler ends than the satisfaction of individual ambitions.' They have 'a code of honour – sometimes, like the Hippocratic oath, historically impressive' (30). Flexner concludes that 'unless legal and medical faculties live in the atmosphere of ideals and research, they are simply not university faculties at all' (30).

A physician with scholarly interests and aptitudes is also likely to make a better psychologist or clinician than a colleague who has mastered only the mechanics of medicine. In his memoir *Geography of the Heart*, Fenton Johnson takes offence at the incessant chatter and self-importance of the Kentucky doctor who examines his dying father. In pleasant contrast are the urbanity and tact of the Parisian specialist whose bookshelves are filled with the great classics of French literature as well as with the expected medical texts. Instead of being asked to remove his shirt within minutes of entering the doctor's office, Johnson is engaged in a long conversation about French literature. As a badge of his humanity and learning, the doctor's knowledge of letters is surely a more civilized credential than the medical diplomas lining the office of his Kentucky counterpart. Johnson is charmed and moved by the Parisian doctor, just as Browning's imagination is quietly stirred by Raphael's writing a 'century of sonnets' and by Dante's producing a single etching of an angel.

The advocate of professional education who opposes liberal learning is little better than a sophist who profanes and parodies Socrates by reducing the wisdom of a way of life to a teachable technique. Unfortunately, the more intellectual acumen a medical scientist displays in the laboratory, the less skill she may possess in eliciting information from a patient or the less dexterity she may exhibit in the operating room. Though a few medical doctors have PhDs, it is rare for a Nobel laureate in biophysics or cancer research to be a great surgeon. There is seldom world enough or time. And even if there were, a research scientist should ideally blend the theoretical reach of a physicist with the contemplative habits of a scholar, who must wait, observe, delay gratification, and be daring and patient at the same time. By contrast, a brain surgeon must combine the arcane knowledge of a neurologist with the swift, split-second response of a lifeguard. But a surgeon is more than an armed savage. He may wield a knife. Yet as Flexner says, the use of a scalpel is an 'accident' of his activity. The essence of surgery 'resides in the application of free, unhampered intelligence to the comprehension of ... the problems of disease' (30).

Yeats fears that 'The intellect of man is forced to choose / Perfection of the life, or of the work' ('The Choice,' ll. 1–2). The very act of succeeding at a profession may cut the specialist off from some important part of himself. And that part could destroy the specialist professionally even as it promotes self-growth and makes him a better human being. To avoid Yeats's dilemma and to minimize Newman's concern that the value of professional education to society is in inverse ratio to its value to the individual specialist, lawyers should try (in Oliver Wendell Holmes's words) to 'live greatly' in their 'profession' (quoted by Bok, 1986, 112) and to practise law as if it were an art like writing poetry.

The university equivalent of a caste system is a faculty in which professors of business, law, or medicine function as Brahmins at the top of a hierarchy; where physicists and chemists occupy some middle territory; and where the professors of philosophy and literature, the despised untouchables of academic culture, can be found begging for alms at the base of the pyramid. In a liberal democracy, any caste system of this kind is always moving toward obliteration and so its invidious distinctions are always in a process of decay. But some compromise between a caste system and a classless democracy may be highly expedient. We are half ruined by academic rivalries and competitions for funding. But perhaps we should be wholly ruined without them. Just as the professor of clinical medicine needs to cultivate the cooperation and friendship of the biochemist or the professor of genetics, so the literary historian needs to find in the literary critic a necessary adversary and living reminder that all history, including literary history, tends to suppress critical intelligence. In an academic culture, as in society itself, it is important to keep our friendships in repair. But we also need our enemies. As T.S. Eliot says, 'within limits, the friction, not only between individuals but between groups, seems ... quite necessary for civilisation. The universality of irritation is the best assurance of peace' (1948, 59). The most productive academic departments, professional as well as liberal, are neither war zones nor social clubs but places of friendly rivalry where a measure of amiable conflict and even controlled envy may promote research and advance higher learning.

An important condition of progress in a practical science like engineering or medicine is that aircraft designers and doctors be free to understand the phenomena involved, 'indifferent,' as Flexner says, 'to the effect and use of truth' (15). Chemistry could not advance until alchemists abandoned their commercial interest in converting base metals into gold. 'So again,' in Flexner's words, 'medicine stood almost still until the pre-clinical sciences were differentiated and set free – free to develop without regard to use and practice' (14). Instead of giving society what it thinks it wants, a responsible faculty of practical science should give society what it needs.

The great medical, law, and engineering schools have to keep reminding the university that it is neither a cloister nor an archive. Each metaphor is too reclusive and sequestered. Though the academy is a place of refuge where the mathematician and poet are free to enjoy disinterested play, it is also a place where disciplined minds and educated imaginations are learning to serve society more responsibly. Despite what some professional schools may say to the contrary, the soul of education is non-vocational. As even Francis Bacon, the herald of the scientific revolution in England, concedes, 'the very beholding of the light is itself a more excellent and fairer thing than all the uses of it; so assuredly the very contemplation of things, as they are, without superstition or imposture, error or confusion, is in itself more worthy than all the fruits of invention' (539). For this reason alone a university that wants to be more useful to society than a museum of mummified theologies or confiscated myths of concern will also have to be more contemplative and more liberal than a seminary or a training school. From one generation to the next the stories a culture wants to transmit have to be patiently rehearsed and retold. And its battles in the darkling plain, its great war against the scourges of disease, poverty, and terrorism, have to be waged anew in each age.

Theology may be the queen of the sciences, but it cannot teach theologians to be holy. Nor can ethics instruct professors to be moral or sociology provide blueprints for Utopia. Too often the humanist's wisdom is a foghorn: in drawing attention to the danger, it does nothing to dispel it. The advantage of professional education is that its hero is Prometheus: with the fire he steals from heaven Prometheus can disperse the fog in Plato's cave. Instead of remaining aloof from the cave, where most lives are spent in chasing shadows, the graduate of the great professional schools of medicine and law can be a sage and humanist, a physician to all men. In their workshops of science and culture, maps of the human genome and models of a just society are continually being redrawn in an effort to unite utopian imagination with a strong sense of fact.

According to Joyce's Mr Duffy in 'A Painful Case,' 'love between man and man is impossible because there must not be sexual intercourse and friendship between man and woman is impossible because there must be sexual intercourse' (Joyce, 112). Though Mr Duffy's epigram fences *out* more meaning than it fences *in*, the dilemma he describes pervades our culture. In our vocations, as in our personal lives, we confront either a love affair that is friendless or a friendship that is unloving. The same dissociation of love and desire afflicts the modern university. As professors and students alike, we desire the technology that allows us to master the mechanics of engineering, law, and medical science. But unless we can first make love to practical knowledge and its complex technologies, as we make love to a poem that fires our imagination or a philosophy that stirs our mind, we shall find it uninspiring and soulless. We love the arts of music and painting, in which we find part of our own life story written. But until we acquire some mastery of their great informing models, until as art historians or practitioners we acquire some discipline and technique, our love remains unconsummated.

At their best the professions are neither cold nor arrogant but wear a 'human face,' like Blake's Pity in 'The Divine Image.' The ultimate civility of medicine is the humane art of young doctors hovering over their patients with acrobatic grace, like crusaders or knights templar in gallant white coats. In the dedicated medical student we glimpse the future physician, the noblest of professionals, whose wholeness of spirit gives his patients strength, whose knowledge gives them hope, and whose greatest gift is love.

7

Personal Knowledge: The Lifeblood of Learning

Without ceasing to be objective and disinterested, higher learning must satisfy the desire of an intelligent human being to be an architect and poet, a fashioner of values and a discoverer of worlds. Scholars and scientists can be extraordinarily creative and resourceful, not because they are supercomputers, but because their ability to integrate reason and imagination allows them to connect their past to a meaningful future. They know that imagination is the source of meaning, which is the antecedent condition of truth. The opposite of meaning is not logical contradiction or simple wrongheadedness but chaos or anarchy, a 'darkling plain' from which personal knowledge and value have ebbed away like the tide in 'Dover Beach.' But even in constructing liveable worlds the mind can dwell in, there is no reason why disciplines such as literature, philosophy, or history need insult a refined intelligence by 'weaving' what Keats calls a fanatic's 'dream,' or a 'paradise for a sect.'

In this chapter and the next I shall argue that the soul of higher learning is a personal appropriation of models and paradigms. Unless the great organizing models of scholarship and science are appropriated as a living and personal possession, no knowledge is possible. And unless an inert body of facts is animated by an informing principle or model, the knowledge is not personal. Models without personal knowledge are empty, and knowledge without models is blind.

Unlike a 'belief *that*,' which is built on a bedrock of proof, a proposal or a 'belief *in*' is a wager. Like all sane risks or wagers, it floats on a raft of primitive assents: logical coercion is never its goal. This chapter tries to float two such wagers or 'beliefs *in*.' The first wager invites a society driven and enslaved by technological desires to embrace the proposition that, without being merely subjective, higher learning can be personal – a 'belief *in*' the value of science or humane learning, for example. My second wager, which is equally personal (and, I hope, equally objective or disinterested), invites assent to the axiom that knowledge can achieve the dignity of a shared consensus – a 'belief *that*' a specific gene can be located in a particular chromosome, for instance – without making all learning conform to the impersonal sequencing, say, of the human genome in DNA technology.

A technological society has little use for speculative 'beliefs-*in*.' It discourages proposals about the structure of cells, atoms, or Shakespearean sonnets, unless these proposals result in the invention and application of new techniques. But just as Hegel predicts that art will someday immolate itself on the altar of dialectic, dying in its most refined form into a purer discipline, so the more efficient technologies become, the more ruthlessly they destroy science, whose chief and only begetter is the mind's capacity to frame speculative proposals. As Browning's Cleon wryly reflects, 'most progress is most failure' ('Cleon,' l. 272). In creating a culture of masters and slaves, technology also atrophies the speculative capacities that allow a master self to rule. The result is that the slave enslaves the master, who becomes (in Thoreau's chilling phrase) the tool of his tools and 'the slave-driver of [himself]' (649).

In his celebrated portrayal of masters and slaves in the *Phenomenology of Mind*, Hegel shows how the bondsman realizes his selfhood through the labour he performs. When Hegel's model is applied to the university, the master turns into the academic administrator paid to hire and evaluate professor slaves. But only professors who do the work of scholarship and science possess the knowledge that Bacon associates with power. Just as a minister of education or the dean of a faculty is seldom indispensable to a scholar's contemplative activity or to a scientist's talent for discovery, so the educated physicist or philosopher may find he is no longer essential to the practical intelligence of the technicians, who acquire the sinister power of the machines in Butler's *Erewhon*.

The genuine selfhood that Hegel withholds from the masters is conferred on the slaves, whose work (or expense of energy with a further end in view) is commendably productive. By contrast, in their expense of energy for its own sake, the educated masters become as purposeless and idle as the drones in a hive of worker bees. Hegel's masters are failed selves because they have avoided the full risk of feeling dehumanized by struggling to find productive work. Though computer technology was once essential to the scientists who invented it, the masters themselves are no longer essential to the computers and technicians trained to operate the hardware, which takes on a menacing shadow-life of its own. Hegel's metaphor of the master-slave captures the paradox of the master who is enslaved by the bondsmen who should guarantee his freedom. Whereas the master wants to dwell in a world of ends freed of means, the slave lives in a world of means without ends.

Hegel is right to identify technology as the servant or slave self that has ironically become our master. But instead of blaming the master's loss of an authentic selfhood on the scientist's talent for discovery or on the humanist's contemplative gifts, we should all gratefully acknowledge that the master self has an important function to perform in fashioning worlds that our imaginations can live in and our intellects explore. As Phillip Verene says, 'the technological world is dehumanizing because there is no technique of the human, of being human. There is no technique for self-knowledge, the knowledge of the self as self, which is the basis of civil wisdom' (150). Hegel degrades this personal knowledge below its innate dignity and worth by asserting it is available to everyone. He fails to see that selfhood is the virtue of the serving master or master servant alone, as he struggles to make ends and means cohere. Until work and play are one, an important task of the educated humanist and scientist is to assist the slave self in its panic-stricken quest to fill with meaning a 'reality that is all surface' and that 'has gambled all on one phenomenon: technique' (Verene, 147, 151).

To say that knowledge is both disinterested and personal may sound contradictory. But unless knowledge is disinterested it will not be objective, and unless it is personal it will not be animated by the pervasive, often unconscious values that ensure it will be meaningful. If a learner were merely to project his personal desires he would soon founder in prejudice. All learning must commit itself, in Stuart Hampshire's words, 'to patience, skepticism, slowness, to minute attention to detail, and to the usual disappointment of large designs' (40). Accuracy and painstaking attention to detail are hard to sustain, however, unless the desire to know is personal and presupposes a prior commitment to the inherent value of the inquiry. Personal knowledge is the precious lifeblood of learning, its breath and finer spirit, because it is the source of meaning. The antithesis of this knowledge is not falsehood or error but the nightmare surrealism of a world in which ideas are only motions of matter or electrical impulses in the brain. No scholar ever believed he lived in such a world. If he did, such words as 'meaning' and 'intention' would have to be exorcised from his vocabulary as relics of an obsolete psychology. Loss of commitment to a world that is intelligibly designed or to an academic project's intrinsic worth is like a loss of religious faith.

A subtle and generous scholar who spends a lifetime of intellectual energy and talent on an inquiry hopes that his 'academic instincts' (to borrow Marjorie Garber's phrase) will be justified by rewarding results or enhancing discoveries. When Henry James's Winterbourne sees Daisy at the Colosseum in the company of Giovanelli, his opinion of her suddenly plummets. We are told 'he felt angry with himself that he had bothered so much about the right way of regarding Miss Daisy Miller' (148). As the subject of prolonged appreciative inquiry in a short story appropriately subtitled 'a study,' Daisy ought to live up to Winterbourne's enhancing imagination of her. Like a scholar or scientist who loses faith in a research project, Winterbourne feels momentarily betrayed by his materials. As Hampshire concludes, 'the really difficult issue of commitment, and of the morality of scholarship, is this: how are we to decide what questions are worth asking, what problems are worth raising, or, more strongly, what problems must be raised? ... We have to presume that scholarship is not merely defensive, a distraction from our first-order concerns, and in this sense detached' (41).

One of the great twentieth-century essays on Homer comes from Simone Weil, whose experience during the Second World War illuminates Goethe's comment, quoted by Matthew Arnold, that from Homer and Polygnotus he every day learns that in our life above earth we have to enact hell. Weil's wartime experience makes Goethe's hell one of her first-order concerns. As a French saint who disturbs believers by self-destructive attempts to imitate a sacrificial God, she is a 'committed' scholar in the sense Stuart Hampshire defines in his comments on the historian Lewis Namier: her scholarship has 'a traceable common base in deeply felt personal needs' (Hampshire, 41). The remorseless cruelty of the *Iliad*, which is in profoundly uneasy conflict with its heroic grandeur, engages Simone Weil's imagination without directly projecting her own strong feelings about Christianity and the Holocaust. The mystic who asks God to devour her and give her away as food to afflicted men who lack every kind of nourishment experiences the tragedy of Hector and his ravaged, grieving family as a displaced version of her own conflict. But like all scholars, Weil also 'perceives a countervailing need to respect the complexities of the evidence' and all 'resisting material.' Without such 'tension,' as Hampshire insists, no 'worth-while scholarship' is possible (42). Like the precocious Arthur in *Tom Brown's Schooldays*, one can be faithful to Homer's Greek without being so deficient in chivalry as to 'murder the most beautiful utterances of the most beautiful woman of the old world' (Hughes, 281) as she grieves over Hector's body. And one can be a scholar without ignoring the irresistible charm both Homer and Weil perceive in the wrath of Achilles. It is psychologically devastating and heartbreakingly sad that so noble a victim as Hector should be defeated by a rival momentarily dispossessed of his humanity by a cruel resolve to be avenged for the death of a friend.

As Stephen Greenblatt concedes in the acknowledgments to his book *Marvelous Possessions*, his criticism of European territorial claims in the New World is also a criticism of the Zionism in which he was raised. In his own words, 'neither the critique nor the bond constitute[s] the meaning of [*Marvelous Possessions*] – which is, after all, about other times and places – but their pressure makes itself felt' (ix). Far from compromising the rigour of Greenblatt's scholarship, his complex ties to Zionism help sustain the intellectual energy and interest that make scholarly precision possible. Similarly, Isaiah Berlin is drawn to the history of ideas, not because of any conscious wish to follow the example of congenial Oxford dons who happen to be philosophers, but because of his intoxicating conversations about Kant, Husserl, and Schopenhauer with a brilliant but slightly mad German exile called Rachmilievitch. Berlin's gift for lightning-swift conversation, like his book on Marx, owes less to Stuart Hampshire and J.L. Austin, whose manners were 'precise, Anglo-Saxon, procedural and rationalistic,' than to his less publicized exchanges with his dark double, the shadowy Rachmilievitch, conspiratorial and foreign, who might have stepped out of Conrad's novel *The Secret Agent*. 'Rambling, historical, saturated in Russian, German and Jewish references,' as Berlin's biographer explains, these talks were 'conducted in three languages at once' (Ignatieff, 81).

Great scholarship or science is never a product of dissociated sensibility. Far from exalting intelligence at the expense of spirit, the genius of Einstein no less than the art of Goethe is an 'imagination that succeeds,' as Santayana says. It is as impossible to separate Spinoza's lofty contemplation of Substance and its modes from his Jewish monotheism as it is to understand the utilitarianism of Bentham and Mill apart from their Anglo-Saxon empiricism, which seems imbibed with their mothers' milk. Whitehead says 'every philosophy is tinged with the coloring of some secret imaginative background, which never emerges explicitly into its trains of reasoning' (quoted by Frye, 1990, 12). Though an unconscious commitment or postulate of faith can degenerate into a mere idiosyncrasy or whim, it can also make essential contributions to a scholar's imaginative background or 'inner citadel,' a phrase Berlin uses to describe personal knowledge. Bertrand Russell's first-hand experience of what it means to live as a lonely orphan in the household of a tyrannical grandmother illuminates the bizarre inhumanity of his admission that he likes 'mathematics because it is not human and has nothing in particular to do with this planet' (quoted by Charlton, 80). More constructively, the grim logic of Calvinism helps explain the corrosive, all-dissolving ironies of David Hume's essays on miracles. The theology of his youth is a grindstone that sharpens the edge of a keenly sceptical mind.

An archaeologist's passionate commitment to exploring a lost Atlantis she discovered in her youth may be as necessary and natural as a mother's love for her child. But to say that God exists because we want him to is like saying a man's opinion is right because he has it. The proper use of personal knowledge becomes an abuse only when, in addition to serving as a catalyst of discovery, it also becomes the essential element of what is found. A biochemist in the pay of a pharmaceutical company may have a vested interest in the results of an experiment to establish the safe use of Thalidomide. And instead of accepting the normal risks of scientific investigation, a naturalist like P.H. Gosse may misinterpret the data. In his book *Omphalos: An Attempt to Untie the Geological Knot* (1857), Gosse's implausible effort to picture God as a manufacturer of misleading evidence makes him a dishonest scientist. In the image of the devious God he invents, Gosse plays the game with marked cards.

It would be wrong to conclude, however, that disinterested learning is dispassionate or devoid of personal feeling. As one commentator says, 'there never yet was a philosopher, ... nor man of science, whose conclusions ran counter to the dearest wishes of his heart, who summed up against them, or condemned his hopes to death' (Dixon, 16). Charles Darwin is merely being more honest than most scientists when he admits there was a time 'when the thought of the eye made [him] cold all over' because it seemed inconsistent with natural selection. 'The sight of a feather in a peacock's tail, whenever I gazed at it, [made] me sick' (quoted by Dixon, 17). Personal experience of random change is not irrelevant to scientific discovery. On the contrary, accidental variation is a key Darwin uses to open nature's lock and wring from its close reserve the secret of evolution. To be sure, a scholar or scientist cannot use untested personal knowledge to assume in his premise what he purports to discover only in his conclusion. But often merely competent research is transformed into great research when submission to stubborn data or resisting evidence combines with commitment to values that are felt to be important. The predictable conclusions of a merely good book like Paley's *Evidences of Christianity* are redeemed by the inductive energies of a great book like Darwin's *Origin of Species*, whose theory seems to emerge spontaneously out of a scrupulous respect for facts.

The career of the Swiss historian Jacob Burckhardt illustrates the importance to learning of civic pride and of allegiance to the values of a special place – a *genius loci*. When Burckhardt refuses to accept the University of Berlin's offer of Von Ranke's chair in history, his decision to remain at Basel dramatizes his commitment to a Swiss ethos and to an appreciative kind of cultural history that the rise of scientific methods in Germany had recently challenged. In Berlin Burckhart might have become another Von Ranke. But writing for his Basel students (who included an admiring Nietzsche) and for the Muse of history herself, Burckhardt lived and died outside the centres of power. As one commentator explains, 'Basel filled Burckhardt with the respect for a cultivated mercantile elite which lights up his account of the Florentine Renaissance and his appreciation of Venice's proud sense of isolation and independence' (Grafton, 53). Realizing that none of the protocols of scientific historiography or 'source criticism' could make the cultural monuments speak, Burckhardt learned to combine a literary critic's ear for nuances of phrasing with a psychoanalyst's knack of retrieving meanings that lay just below the surface of historical narrative. Lecturing ten hours a week without notes, Burckhardt managed to produce books for his friends and students that are still read today, while the tomes (and tombs) of his Berlin contemporaries collect the dust of centuries.

It is a mistake, I think, to assume that scholars typically magnify the importance of attributes they themselves possess. Often they are more likely to pay homage to qualities they lack. No philosopher of science writes more persuasively than Karl Popper about the function of falsification and critical testing in the development of science. But Bryan Magee, one of the few people who knew Popper well, says that Popper 'turned every discussion into the verbal equivalent of a fight, and appeared to become uncontrollable with rage' (198) whenever Magee disagreed with him. Popper welcomed dispassionate criticism in principle precisely because he never acquired a tolerance for it in practice. He was paradoxically hostile to all criticism of his theories, including his cherished principle that all theory should be criticized.

Though in scholarship, as in poetry, energy seems to come from the investigator's unconscious prior commitments, these allegiances have to be appropriately distanced and detached if the exploration is to be rigorous and the analogies exact. Such at least is John Keble's proto-Freudian theory of poetic creation, according to which all epic and dramatic genres displace a poet's lyric impulses and personal commitments. If Virgil is composing a pastoral poem or an elegy, he is writing under little restraint to 'any sudden burst of high or plaintive feeling' in which he may choose to indulge. Such is not the case, however, with classical epic, in which Virgil may have to displace his instinctive 'love of woods and rivers' by forcing that instinct to work 'its way through all the incumbrance of the epic story.' 'Interposed, as a kind of transparent veil, between the listener and the narrator's real drift and feelings,' both the narrative structure of a prose romance by Sir Walter Scott and the epic action of the *Aeneid* function 'like a safety-valve to a full mind' (Keble, 440, 436). The only tolerable personal commitments, in art as in scholarship, are a product of the distancing that keeps them appropriately disguised and detached.

Imaginative energy has its roots in individual biographies. But until the source of this energy is displaced it can seldom receive the scrupulous channelling and refinement it requires. In canto 4 of the second part of Tennyson's *Maud*, the lament 'O that 'twere possible' (2.141–238), a lyric of loss originally composed in 1837 for Arthur Hallam, reveals how a whole narrative poem, like a pearl grown from a grain of sand, has organized itself unconsciously around a forgotten fragment.

> O that 'twere possible
> After long grief and pain
> To find the arms of my true love
> Round me once again!
>
> When I was wont to meet her
> In the silent woody places
> By the home that gave me birth,
> We stood tranced in long embraces
> Mixt with kisses sweeter, sweeter
> Than anything on earth.
>
> *Maud*, 2.4.141–50

The association of this buried fragment with the frugal tetrameter quatrains of *In Memoriam* discloses in a flash that the speaker's love for

Maud has its origin, not just in Tennyson's unrequited love for Rosa Baring, but also in his deeper love for Hallam, whose gender has been switched in the poet's narrative displacement of a hidden memory.

If Tennyson had expressed his love more directly, it would have been charged with much less energy and power. For often the truths we find the words to speak have already died in the heart. I doubt that Robert Frost would have claimed Arnold's 'Cadmus and Harmonia' was his favourite poem if its catastrophes had not unconsciously prefigured his own domestic tragedies. Even when Frost's favourite daughter Marjorie dies and takes his heart 'out of the world with her,' the tragedy of that loss finds its objective counterpart and partial release in Arnold's lyric. As William H. Pritchard says, 'what is so extraordinary, so heroic about [Frost's] letter [to Louis Untermeyer], is that in the face of the most awful, senseless, heart-piercing loss – the death of a child, ... Frost refuses to give up poetry also. Indeed, the culminating moment of his account ... of the event is the allusion to Arnold's poem' (195).

The same law of indirection applies to scholarship. To aspire too directly to political or social relevance is to lose it. Though important life-and-death issues have to be raised, they should seldom be explored explicitly or head-on. In Stuart Hampshire's words, 'imaginative energy has largely incalculable sources and serves largely unconscious needs. The only safe criterion is the degree of intellectual excitement that a work or a problem – in, say, philosophy or literary criticism – provokes ... To put it paradoxically: in this field one should do, as a matter of policy, what one strongly feels that one must do, without any policy at all' (55).

Since knowledge can be personal without being subjective, a committed scholarship or science should be as attainable in principle as a thought that passes instantaneously into an emotion or as 'a feeling that flashes back,' in George Eliot's phrase, 'as a new organ of knowledge' (166). Like Stuart Hampshire, Judith Shklar believes that the best committed scholarship has unconscious personal origins. Only the energy of an obscure passion for ideas can engage the scholar's interest in a problem over long periods of time. The best scholarship and teaching are 'a complex response to a primary passion. Something has set it off, possibly an ideology to which one is committed or its collapse, or an historical experience, such as the Second World War lived in Europe or the Depression in America, or the felt need to find new political expressions for an altered and only half understood political world' (Shklar, 152). Without the involuntary diffusion of such energy, the dullness of academic routines and the expense of life and mind in a youth ghetto would soon become intolerable for most instructors.

Only an abiding personal commitment to a problem in ethics, genetics, or biochemistry allows an aging scientist or scholar to offer her 'graduate students a genuine picture of what real thinking is like.' Too often, Shklar predicts, 'tedium, indifference, and stagnation [become] the typical afflictions of older teachers' (157). But if scholarship without commitment is dull and unimaginative, enthusiasm without self-scrutiny is mere showmanship and theatre. The academic charlatan usually assumes one of two disguises: he may masquerade as a pedant whose learning is too uninspired or timid to appeal to anyone. Alternatively, he may set up as a charismatic guru whose strong personality and lust for power mask a thinly disguised contempt for disciplined thought. Celebrity-seekers cheat their students because, as Shklar shrewdly observes, 'they are basically interested in themselves, in *their* message, in *their* followers, and in *their* renown' (158).

Whereas the guru is more absorbed by his own charisma and learning than by the welfare of his students, a true Socratic mentor protects his disciples against personal influence. His authority is self-dissolving. In a prophet of genius every potential disciple finds a reflection of his own rejected thoughts. The prophet reminds his students that, as lamps rather than mirrors, cloning is unworthy of them. There is always room for charismatic prophets like Harold Bloom or Oscar Wilde. But as Bloom himself admits, such oracles may be too deficient in negative capability to rank among the greatest teachers. Quoting Wilde's maxim that 'one should never listen. To listen is a sign of indifference to one's hearers,' Bloom adds wistfully: 'I haven't won any teaching awards in my half-century career, because I believe in the passion and intellect embedded in that apothegm' (2002, 246).

Just as the sophist as prophet or guru offers a mere parody of the scholar's personal knowledge and commitment, so the sophist as pedant profanes the impersonal authority of an unbiased thinker. Though detachment and commitment are both necessary, David Perkins accurately observes that the importance ascribed to each of these contradictory virtues may vary from one generation to the next. In the 1950s Perkins believed that 'the cardinal point of classroom morality' was 'to present literature impersonally and objectively' (1988, 112). Under the sway of T.S. Eliot's preference for significant rather than sincere emotion, Perkins originally thought that 'the teacher-critic was analogous to

[Eliot's] impersonal poet' (1988, 113). Thirty years later he reverses these priorities by ranking personal knowledge above scholarly detachment. Influenced by Hans-Georg Gadamer's theory of a textual 'encounter' that 'speaks to our deepest concerns,' Perkins later concludes that 'since all reading and understanding are inevitably personal, it is phony to pretend otherwise. In fact, a personal interpretation claims less authority and promotes dialogue better than an impersonal one' which claims the assent of every unprejudiced reader (1988, 113).

To beget genuine passion in his students, a teacher must love the subject he teaches. But to love and care for a subject is also to take care to understand and master it. As John Passmore observes, 'a love without care is not love at all' (1980, 195). Careless love is 'pure play, just as love without enjoyment is pure toil.' A genuine intellectual love 'involves understanding, imaginative thinking, consideration, courage, a willingness to work and to exercise one's intelligence' (Passmore, 1980, 195). To maintain, however, that all personal knowledge is subjective and therefore unscholarly is to assimilate learning to a positivist model that is dangerous and illusory.

Such at least is Isaiah Berlin's argument in his essay 'Historical Inevitability.' 'Except on the assumption that history must deal with human beings purely as material objects in space – must, in short, be behaviourist – its method can scarcely be assimilated to the standards of an exact natural science' (165). To posit a 'purely descriptive, wholly depersonalized' model is to distort rather than clarify historical assessments (166). Literary critics are no less meticulous in plotting the moral geometry of novels by George Eliot and Henry James than Euclid is in demonstrating his theorems. But since there are degrees of precision appropriate to each art and science, the criteria will be different. Indeed Berlin believes that to make an idol of mathematics or science by renouncing personal knowledge in historical research is 'to practice austerities, and commit deliberate acts of self-laceration' which 'render the writing of history (and, it may added, of sociology) gratuitously sterile' (165).

Hypocrisy, by contrast, is as great a vice in pedagogy as it is in ethics. To apply for the Mammon Professorship of God is to play the role of an imposter and opportunist. It is to simulate a passion or love for theology one does not possess. To commit an act of bad faith or unconscious self-deception is a profane caricature of scholarly detachment. It is just as unworthy as a guru's anxiety to maintain his star status by parodying a prophet's personal values and commitments.

Two of the subtlest advocates of personal knowledge in higher learning are the philosopher Francis Sparshott and the chemist and philosopher of science Michael Polanyi, author of a book entitled *Personal Knowledge*. In his essay 'Credo ut intelligam' Sparshott argues that the function of an academic lecture is to impart personal truth. 'I cannot,' he confesses, 'bring myself to say in the lecture hall anything whose truth does not at the moment seem to me a matter of interest and possible concern.' Though the lecturer should make notes beforehand, he should use them 'only as insurance and control.' In the lecture itself professors 'should compose [their] souls to silence and allow whatever may be in them to well forth – not "We lecture" but "It lectures in us"' (1972, 40).

Ideally, Sparshott's lecturer in ancient philosophy is the oracle through whom Plato or Aristotle speaks: the Muse dictates the lecture and the professor is her scribe. Such a lecturer combines the prophet's gift for hearing voices with the ventriloquist's gift for projecting them. Sparshott has little advice for lecturers who try faithfully to climb Parnassus every Friday afternoon only to find that the gods are on vacation or the Muse is out to lunch. In such a crisis the lecturer should presumably fall back on cue cards and notes. Sparshott's model, however, is not Plato's Ion or even John Milton, who claims that the true author of *Paradise Lost* is the Holy Spirit. The most authentic prototype of Sparshott's lecturer is Alfred North Whitehead, who says that 'so far as the mere imparting of information is concerned, no university has had any justification for existence since the popularisation of printing in the fifteenth century' (1949, 97). Only personal knowledge imaginatively imparted can provide relief from what Whitehead calls 'the horrible burden of inert ideas.' The university 'imparts information, but it imparts it imaginatively' (1949, 13, 97).

A professor's lectures are to his written books what these books themselves are to the textbooks of his discipline. Though the humanity of a lecture may get in the way of a lucid outline, the spectacle of a scholar thinking is less likely than a passive reading of a textbook to soften a student's brain or inoculate him against any future outbreaks of intellectual zeal. Similarly, a delicate and committed treatment of a topic in a scholarly monograph always explores more interesting terrain than the no-man's land of an encyclopedia article or an entry in a handbook. Sparshott concedes that 'first steps are bound to be dull, and textbooks bore in a good cause' (1972, 41). But it is fatal, he insists, to mistake 'slogans and summaries' for the thoughts they stand for. Though Hegel's aphorism that 'the truth is the whole' is a useful reminder of the experience of reading his *Philosophy of History* or even his more stirring *Phenom-enology of Mind*, which loads every rift with ore, we abuse its shorthand if we substitute his slogan for a bold immersion in the tidal surge and flow of the treatises themselves.[1]

Sparshott contrasts the textbook or survey of a subject, which he calls a 'starting-point' or a 'way-station,' with the 'destination' that provides not just an informing idea or an end-point but also a context for under-standing the subject as a whole (1972, 41). Sometimes to grasp an expe-rience instantaneously and as a unit, we have to collapse all words into phrases of equivalence, as Hopkins does in his dazzling perception of a raptor in flight.

> I caught this morning morning's minion, king-
> dom of daylight's dauphin, dapple-dawn-drawn Falcon, in his
>   riding
> Of the rolling level underneath him steady air
>   'The Windhover,' ll. 1–3

Such language is electric with energy. It does not allow us to sort out atomic impressions of dapple colour, scintillating light, and aerial motion. It communicates instead, in a single mounting impression, a parataxis of highly charged sensations: a princelike bird drawn by a char-iot of dappled light; a spacious riding-of-the-rolling-level-underneath-him-steady-air.

If the 'truth is the whole,' then our estimate of Hegel will depend less on the truth or falsehood of individual propositions than on what Susanne Langer calls his philosophy's 'presentational form.' Like Hop-kins's windhover, this form is something that must be apprehended all at once or not at all. Indeed Hegel's unifying vision of the world may not be stated discursively in any single sentence of his works. The secret of presentational language is its truth of embodiment, which philosophy and art sometimes offer as an alternative to truths of logical demonstra-tion and rebuttal. The great philosopher and artist live questions they cannot answer. Often they cannot know the truths they embody or artic-ulate the meanings they exhibit.

Sparshott's distinction between 'way-stations' and 'destinations' antic-ipates an important distinction between 'maps' and 'models' which I explore in the next chapter. It is enough to observe here that mastery of a 'model's' validating context can best be transmitted to a student who is working alongside a mentor. Sometimes such an apprentice can see in a flash the insights or convictions that are otherwise incommunicable and that can seldom be imparted directly by even the most gifted lecturer. For this reason, the better we come to know and live with a philosopher like Hegel, the less we care whether he is right or wrong. As Sparshott says, 'what is true or false is not the idea as it is believed in, but the lifeless doctrine' of a mere survey map or handbook entry 'that can be abstracted from it' (1972, 45).

Students attend lectures and read Hegel himself rather than some mere textbook summary of his thought because in order to understand the 'presentational form' of his philosophy they have to eavesdrop on 'the sheer human worth of a man delicately and passionately thinking just so' (Sparshott, 1972, 43). Unlike the cardboard cutouts of a survey course, in which bloodless ideas flit like phantoms across the stage of history in a game of intellectual charades, a study of models shows how any genuine drama of ideas instructs by example. No distance learning on the Internet can take the place of live seminars or lectures. Only the give-and-take of animated discussion can provide the presentational equivalent of Hopkins's self-embedding Latin syntax ('Of the rolling level underneath him steady air'), which sunders the preposition 'of' from the direct object 'air' and all but dissolves the windhover in undulating light. Just as no paraphrase can provide a substitute for Hopkins's collapse of language into one electrically charged and untranslatable synonym, so no textbook can convey the challenging three-dimensional depth of a lively seminar on Hopkins's poetic grammar. Nor can it simulate the stereoscopic vision of Hegel's *Phenomenology of Mind*, where a range of insights seems endlessly to recede, as 'Hills peep o'er hills and Alps o'er Alps arise' (Pope, 'An Essay on Criticism,' 2.232).

The scholar's commitment to personal knowledge involves the exercise of at least three different virtues. If one must first believe to understand, as Augustine insists ('Credo ut intelligam'), then personal knowledge posits the value of lived experience, since faith in such knowledge is a precondition of all learning that embodies what it knows. One might even risk the generalization that the ultimate postulate of faith is the rationality of the universe. Unless the scientist can posit without proof the uniformity of nature, and unless the scholar can inhabit an intelligible world, no laws can be discovered and no meanings explored.

The alternative to faith is Dante's and Hopkins's more logical assumption that love of God must wait upon knowledge of God: 'I greet him the days I meet him, and bless when I understand' (Hopkins, *The Wreck of the*

Deutschland, 5.8). Unfortunately, such a blessing may be long deferred. For to understand God is to stand under a being who has just been described as standing 'under the world's splendour and wonder' (5.6). As a poet of theological paradox, Hopkins knows that to understand the very principle of 'standing under' is to share Goethe's 'shudder of awe' in a free fall through space. It is to grow dizzy with the prospect of an infinite descent.

As Sparshott observes, however, 'faith is not the only theological virtue: there is also love' (1972, 46). Without sympathetic imagination, which entails empathy for distant points of view, a scholar would know only his own system of thought. Sooner or later all his ideas would become dead or inert: he would even lose personal knowledge of his own favoured axioms. Finally, without hope, which is the third theological virtue, few learners would ever commit themselves to the hazards of trailblazing. To follow knowledge 'like a sinking star' is to risk falling off the edge of the world 'beyond the utmost bound of human thought.' But instead of being shipwrecked on purgatory like Dante's Ulysses, an explorer who endures to the end may reach the South Pole and become a hero like Richard Byrd. The satisfaction of completing a difficult but worthwhile project is often in direct ratio to the risk of failing at it.

The most inclusive defence of personal knowledge comes from a distinguished chemist who is also a philosopher of science, Michael Polanyi. Though Polanyi would endorse most of Sparshott's arguments in defence of personal knowledge in 'Credo ut intelligam,' they disagree on one important issue. Sparshott contends that 'the possibility of science depends upon its propositions being treated as abstractions. The propositions which go to form a scientific theory do indeed depend upon a context: ... they take their meaning and importance from the theory of which they form a part, but the theory itself is supposed to be public in the sense that it means the same to all who understand it' (1972, 45). In other words, scientific theories are textbook abstractions, purified of the risks and commitments that often assist in their discovery. To the degree that science is an abstraction, it is not a form of personal knowledge. And to the degree it is personal knowledge, it may be a history of physics or a chemist's biography but never a theory of science as such.

By contrast, Polanyi's book *Personal Knowledge*, a sustained polemic against abstraction, would take Sparshott to task, I think, for offering too detached a picture. Central to Polanyi's argument is the understanding that personal knowledge is not subjective but a form of intellectually committed learning with its own appropriate criteria of what it means to be objective. Personal knowledge is imaginatively committed because its proposals are the antecedent condition of truth: their antithesis is not simple error but an incoherent welter of subjective impressions that make meaningful discovery impossible. Personal knowledge is also liberating because it denies that intellectual activity is like a robot capable of conscripting scientists or scholars into an army of worker ants or drones. An active intelligence is powerless to enslave learners to soulless technologies or to operations that are anti-creative and mechanically predictable.

Any impersonal use of words or numbers according to strict rules is merely a 'routine performance' because, as Polanyi explains, 'it leaves its interpretative framework – the numbers used in counting – quite unchanged.' By contrast, personal knowledge modifies its interpretive context by using an original metaphor or a new mathematical notation 'to modify' the act of interpretation itself (105). To invent a metaphor is to cross a boundary into *terra incognita*. When Robert Frost risks the perception that 'nature's first green is gold,' he takes 'a plunge,' as Polanyi would say, but only to 'gain a firmer foothold' in such flashes of insight as 'Eden sank to grief' or 'Nothing gold can stay.' To discover a metaphor and let it 'run a course of lucky events' is to begin 'in delight' and end sometimes 'in wisdom' (Frost, 1968, 18). Such education by poetry is the paradigm of all personal knowledge. For as Polanyi discerns, its stay against confusion 'is a tacit, irreversible, heuristic feat; it is a transformation of our intellectual life, originating in our own desire for greater clarity and coherence, and yet sustained by the hope of coming into closer touch with reality' (106).

Because there are no rules for inventing metaphors or having insights, 'the logic' of personal knowledge 'is backward' or 'in retrospect,' as Frost says, 'after the act.' Like the slave boy's discovery in Plato's *Meno* that his knowledge of Pythagoras's theorem is a form of reminiscence, any recovery of knowledge we once possessed but then lost is a drama that progresses 'from delight to wisdom.' Knowledge that is personal because we once discovered it for ourselves, and can relive that discovery when we choose, never loses 'its sense of a meaning that once unfolded by surprise as it went' (Frost, 1968, 20).

Polanyi assails as the ultimate sophistry Laplace's mechanical, third-person view of the world, which culminates in his infamous boast of being able to dispense with the hypothesis of God. If Laplace's future is predictable and devoid of risk, it is only because he commits a fallacy of incomplete induction by substituting for a mere 'knowledge of all atomic data' a vastly more inclusive 'knowledge of all experience' (Polanyi, 141). In the Laplacean delusion that all genuine knowledge is rigorously impersonal, Polanyi denounces a misguided conviction that threatens to destroy the mind's greatest achievement, science itself. For whenever the utility of science in 'strengthening the public power and improving the standard of living' is valued more highly than 'the intellectual passions of scientists' themselves, 'scientific value will' sooner or later 'be discredited and its appreciation suppressed' (Polanyi, 142).

If biochemical man is merely the sum of his constituent DNA, how could he ever have discovered the chemical base points of DNA? Indeed, if scientific knowledge is as impersonal as Laplace claims, science itself could not exist. For any consistent third-person view of the world reaches the zenith of a magnificent absurdity. It is as if the world were to become a great theatre in which the best plays of Sophocles and Shakespeare are lavishly produced and faultlessly performed to an empty house. Or as Polanyi puts it, the absurdity comes from 'a passion for achieving absolutely impersonal knowledge which, being unable to recognize any persons, presents us with a picture of the universe in which we ourselves are absent. In such a universe there is no one capable of creating and upholding scientific values; hence there is no science' (142).

To be original is to be passionately committed, not impersonal. For without a passionate desire to achieve something new, nothing risky or innovative would ever be attempted. As Polanyi explains, we call personal knowledge 'creative because it changes the world as we see it, by deepening our understanding of it. The change is irrevocable ... Having made a discovery, I shall never see the world again as before' (143). To the extent that the scientist is a Wordsworthian observer who half-creates 'the mighty world of eye and ear,' Newton and Einstein, no less than Shakespeare, are architects and fashioners of the universes they see. Unlike an analytic proposition in which the predicate repeats the subject, Newton's and Einstein's theories are synthetic *a priori*. They tell us something important about the scientist's own intellect, values, and 'truth-bearing passion.' After a scientific revolution, our 'eyes have become different,' Polanyi says; 'I have made myself into a person seeing and thinking differently. I have crossed a gap, the heuristic gap which lies between problem and discovery' (143).

Since all learning assumes the truth of unproven axioms, it is better to acknowledge and test these axioms than conceal them in a science that is 'blind and deceptive' (Polanyi, 268). Indeed faith in a scientific model can sometimes be just as revolutionary as scepticism or doubt. Max von Laue, for example, discovered the diffraction of X-rays by crystals through 'the sheer power of believing more concretely than anyone else in the accepted theory of crystals and X-rays' (Polanyi, 277). And though Vesalius is rightly 'praised as a hero of scientific skepticism for boldly rejecting the traditional doctrine that the dividing wall of the heart was pierced by invisible passages,' Polanyi admits with wry amusement that 'Harvey is acclaimed for the very opposite reason, namely for boldly assuming the presence of invisible passages connecting the arteries with the veins' (277). Since faith in a model can be just as instrumental as scepticism in promoting discovery, the only rule for entertaining a valid hypothesis is that there is no rule.

Far from providing a crucible of truth, doubt may be as blindly dogmatic as faith. Indeed 'to destroy all belief would be to deny all truth' (Polanyi, 286), since only within a 'structure of commitment' that honours and tests beliefs is personal knowledge possible. Committed thinking takes place only in a zone of free activity with clearly defined upper and lower limits. At the lower limit is a threshold of painful experience like hunger or grief to which the mind passively and unwillingly submits. It makes no sense to commit one's mind to an animal appetite or a disabling illness. At the upper limit of the scale Polanyi places 'forms of intelligence in which our personal participation tends to be reduced for quite different reasons' (301). In a computer designed to simulate a skilful chess player there is no opportunity either to invent new moves or make mistakes. In realizing the 'ideal of a completely formalized intelligence,' this upper limit 'would eliminate from its manifestations every trace of personal commitment' (Polanyi, 301). For this reason alone scientists and scholars should never aspire to the condition of a computer program or a robot wired for artificial intelligence. Such a mind is anti-creative and without the power to originate or invent.

Paradoxically, the efficient application of technology in the prevention of disease or the solution of a problem in formal logic is like the operation of a well-designed computer program. It may take a genius to invent the computer and a medical pioneer like Jonas Salk to discover a vaccine for polio. But once the invention is made, even the dullest technician can operate the program or administer the vaccine. Though refinement of technology is a consummation devoutly to be wished, creative invention tends to abolish itself in the very act of reaching its goal. Unless a scientist is free to make mistakes in ways that a computer is never free to do, he is not at liberty to make discoveries either. Milton insists that to be rational is to be free: though Adam and Eve are strong enough to have withstood Satan, they are 'free to fall.' Without the freedom to give wrong answers and make bad choices, J.S. Mill believes we would lack the dignity of rational creatures. A self-respecting scholar or scientist never forfeits the opportunity of convincing opponents he is right by avoiding the risk of being refuted or proved wrong.

The mind is free to be an architect of ends and a designer of means to achieve these ends only in an enhancing middle region. This distinctively human zone lies somewhere between two extremes of enslavement: bondage to human appetites (at the lower end of the scale) and enslavement to the formalized intelligence of logic or mathematics at the upper end. As a product of personal knowledge or commitment, discovery may often be 'impassioned, sometimes to the point of obsessiveness' (Polanyi, 301). But to reject the intellectual obsessions that allow a scholar to appropriate knowledge as a living possession is illegitimately to 'secede from the commitment situation' (Polanyi, 304). To assert the essential sanity of a scientific or poetic genius is to assert that his personal knowledge is impersonally given. Though great wits may seem closely allied to madness, the difference between a genius and a madman is that the genius is a madman who succeeds, a visionary who guesses an important principle of experience. It is true that no one can read Sylvia Nasar's moving biography of John Nash, *A Beautiful Mind*, without being shocked by the many mathematical geniuses who, either at the height of their powers or in later life, suffered devastating mental illnesses. But what distinguishes, not merely knowledge from information, but wisdom itself from knowledge, is the mind's bold appropriation of the data. As Oliver Wendell Holmes explains, 'three-story men idealize, imagine, and predict. Their best illuminations come from above through the skylight' (quoted by Charlton, 33) without any guarantee that what they see is always enhancing or even sane. The opposite of wise illumination is insanity: the madness of an unintelligible maze of Internets and websites from which all work stations and personal computers have been removed. The world of the fact-collectors makes no sense until scholars and scientists invent its meaning and interpret it imaginatively.

J.S. Mill maintains that teleology or the science of ends is the function of scholars who tell society what goals it should pursue, though not how to reach them. Unless, however, a sane imagination or intelligence is able to purify a society's desired goals and focus its ends, its myths of concern may degenerate into mere superstition or bigotry. As Newman argues, the best defence against bigotry is a sympathetic imagination. 'Narrow minds have no power of throwing themselves into the minds of others ... they think that any one truth excludes another which is distinct from it, and that every opinion is contrary to their own opinions which is not included in them' (1887, 307–8). In religion, as in poetry, there is no substitute for the intelligence of love or for what Keats calls 'the holiness of the heart's affections' and the truth of the imagination (257).

To pass from one branch of knowledge to another is generally to travel between two systems of belief. The mathematical physicist's understanding can be expressed most efficiently in terms of statistical mechanics, from which his knowledge of thermodynamics is then derived. But usually a transfer from discourse A to discourse B is not logically possible, because B is qualitatively distinct from A and cannot be reduced to it. Concepts such as code and codon, replication and self-organization are required to make molecular biology work. But they are not immediately useful to the biochemist, who speaks a different language. As one observer, David Berlinski, says, 'the philosopher asking innocently for an account of life is hardly in a position to dismiss on principled grounds any number of possible answers – a play of biochemical forces, physics in its most complex state, the coordinative interaction of large and small molecules (Watson's answer), aspects of the Mind of God, the structures forged to protect the gene, the appearance in the universe of pity and terror' (275–6). Once we stand at the centre of a belief system, our personal commitment to that system makes it hard to imagine any world outside it. 'Like Hell itself, which is said to be protected by walls that are seven miles thick, each such system,' Berlinski warns, 'looks especially sturdy from ... inside' (277).

But if entrapment in a belief system like the survival of the fittest constitutes its own version of hell, it is also a proof of barbarism not to be interested in the social Darwinist who builds the system. Unless scholars are acquainted with the personal commitments and values of hero-worshippers like Carlyle and Nietzsche, who discern in history the lengthened shadow of great men, they will not know what to admire and defend in democracy. Without personal knowledge of their opponents, they can have little intellectual sympathy for the architects of their own folkways and myths. In a degrading reduction of professors to cheer-leaders, bureaucratic models too often elevate deans and administrators into a faceless hiring authority of CEOs acting on behalf of corporate sponsors and lobbyists dispensing patronage to unthinking partisans. Even in celebrating the values of a liberal democracy, patriots who cannot refute their critics will be defending something they imperfectly understand. It is only a matter of time before the treatment of mindless conformity as a national resource grows into an octopus that clutches every patriot in its grasp.

Just as religion, according to A.N. Whitehead, is what a person does with his aloneness, so scholarship might be defined as what a thinker does in solitude with his intellectual passions and commitments. Julius Getman maintains that the personal values of scholars who embody what they know distinguish original thinkers from the mere novices or apprentices (45). Without allegiance to personal values, few scholars will ever discover how intellectually rewarding (and lonely) higher learning can be. In addition to committed thinking and solitude, good scholarship 'requires a willingness to risk failure and to take the time necessary to pursue a scholarly vision, both of which are discouraged' by the pressure to be productive and to publish too soon. 'Although the current system has produced much writing in all the disciplines, it has not led to very much useful scholarship' (Getman, 48), which takes years of steady application to test and refine.

Unless learning is passionately sought, it will never qualify as personal knowledge. And unless it is in some sense an education in eros, as Plato understood, it will lack commitment. Too often we fall in love, not with the mystery of genetics or with a puzzle in logic or metaphysics, but with the glamour of becoming a prosperous business executive or a lawyer, or even with the satisfaction of scoring high on an exam. As James Agee stunningly expresses it, schools should teach the student how 'he may hope to guard and increase himself and those whom he touches.' They should also give him some 'indication of the damages which society, money, law, fear and quick belief have set upon these matters and upon all things in human life' that concern him most intimately (quoted in Postman and Weingartner, 46–7).

As Marx foresees, the enemy of personal knowledge is not just consumerism but the division of higher learning into two opposed activities: into meaningful self-formation, on the one hand, and social utility (or the satisfaction of other people's wants), on the other. The second alternative is usually a breeding ground of boredom and despair. From such misshapen progeny of this division as human cloning or the proliferation of biological weapons, biochemists find themselves permanently estranged. And in the mindless production of doctoral dissertations that feed like mosquitoes off the lifeblood of their supervisors, scholars confront the systematic parody or abuse of their own discoveries.

Though scholars often produce new knowledge, this is not their primary function. Instead of imitating scientists, they aspire to forms of *self*-knowledge. As David Bromwich says, scholarship 'is something less than knowledge and something more: the learning of arts and habits' (110). Without these habits, knowledge fails to become as irreplaceably and uniquely personal as 'that taste on the tongue ... of an individual style' which is 'stubbornly itself,' a 'single and unrepeatable combination of elements' (Vendler, 1988a, 20–1). Poking fun at those judicious Solomons who know all about Sheba without ever knowing her carnally (148), Mortimer Kadish believes that a scholar must first be enthralled by an academic discipline and ravished by its warm and breathing beauty before making love to it or even thinking about it critically.

According to J.H. Newman, the best way to acquire education in personal knowledge is to live in a college residence, where scholars may benefit from the conversation of fellows and tutors. Too often the corridors of university power fill up with sophists determined to make academic studies more professional. But colleges remain the more congenial haunt of Socrates, who is committed to the less ambitious but perhaps more worthy task of humanizing knowledge. In Newman's view, a college community provides deep and lasting attachments for its members and a principle of stability to balance the commitment to progress embodied in the university at large. The personal knowledge that is cultivated by a study of literature, the classics, and mathematics is also nurtured by Newman's attachment to the little platoon of scholars and fellows who occupy the halls of his undergraduate college, which he once hoped would be his 'perpetual residence, even unto death' in Oxford (1968, 183).

Whereas the university at large is often a lonely and austere house of intellect, the college is a community of occasional affection and intimacy. On rare occasions a college instructor may even receive (to his surprise) a heart-shaped sucker with a note attached: 'Happy Valentines, Professor Shaw. You're loved, respected and highly esteemed by us all.' For Newman the beauty and intelligence of the Oxford colleges are inseparable from their mixture of high tables and gardens, their private courtyards, athletic games, and intellectual musings. Though his tribute is deeply felt, it may strike us today as a trifle precious. Newman speaks nostalgically about the college as 'a sort of shrine' to which the scholar 'makes continual silent offerings of attachment and devotion. It is a second home, not so tender, but more noble and majestic and author-itative' (Cameron, 42). For my taste there is a little too much talk in Newman about grand buildings and ancient chapels, and his love of tradition verges on fetishism. He cannot abide, it seems, an inferior sermon or an imperfect tea cup.

Though Matthew Arnold disclaims for himself the honourable title of Professor, he is quite prepared to invest the University of Oxford with the same mantle of authority as Newman later invests the church. The dignity of any true source or centre is immune to the vagaries of fash-ion. 'Who would not gladly keep clear, from all these passing clouds, an august institution which was there before they arose, and which will be there when they have blown over?' (1961, 234). In a concluding paean to Oxford, Arnold stoops down to gather up his 'puny warfare against the Philistines' into a nobler, more inclusive cause, a fight which better and truer sons of Oxford will wage long after Arnold and his contempo-raries are gone. The true champion of Oxford's cause is not just Arnold but such illustrious friends of sweetness and light as Schiller and Goethe, and all who try to free us from the bondage 'was uns alle bändgit, DAS GEMEINE!' The German slogan is the banner under which Oxford's 'unworthy son' chooses to enlist. Its English equivalent is Arnold's aphorism about the pursuit of perfection. It echoes Keats's aphorism at the end of 'Ode on a Grecian Urn' ('Beauty is truth, truth beauty') and the fear expressed in 'Lamia' that all charms fly at the 'mere touch of cold Philosophy.' 'Who will deny,' Arnold asks, 'that Oxford, by her ineffable charm, keeps ever calling us nearer to the true goal of all of us, to the ideal, to perfection, – to beauty, in a word, which is only truth seen from another side? – nearer, perhaps, than all the sci-ence of Tübingen' (1961, 236). Perfection is the axis of repose where sweetness and light, the two poles of Hellenism, come together in a sin-gle harmony, where each supplements and sustains the other. If such a Utopia is an unreal paradise, it is nevertheless the fond haven of which Oxford herself, an incorrigible romantic, has always dreamed. Even in distancing himself from Oxford as an 'unworthy son,' Arnold uses the 'home of lost causes' and 'impossible loyalties' (1961, 236) to sanction his own crusade, imperfect as it may be, against the philistines.

T.S. Eliot rightly observes that Arnold's rhapsodic hymn to Oxford, 'spreading her gardens to the moonlight, and whispering from her towers the last enchantments of the Middle Age,' has 'not worn' well (1932a, 448). And yet nobody knows better than Eliot the importance of commitments to 'the life of significant soil.' The art of academic government is the organization of idolatry. Gothic towers and domes are venerated by every college benefactor. Even the conferring of diplomas at convocations, which presume to admit graduates to the society of educated people, is an artificial hallucination in which the presiding chancellor combines 'the inertia of a wooden idol with the credibility of a flesh and blood one' (G.B. Shaw, 1946, 271). I doubt that wearing black gowns to dinner and taking meals at long wooden tables where masters intone grace in Latin are desirable in themselves. But I know that Arnold's battle in the darkling plain – his war against omnipresent philistinism and sophistry – is still waged today. And I strongly suspect that a North American college is more likely to win this war if it rejects the pretension of any ritual suggesting High Tea at Oxford and embraces its own traditions instead. For if history 'is a pattern / Of timeless moments,' a university 'without history / Is not redeemed from time' ('Little Gidding,' 5.20–2) but turned into a cloistered cell or prison of flickering images, like the shadows in Plato's cave.

When people talk about personal knowledge, about their hobbies or their special contribution to a discipline, they are nearly always interesting. Though Newman's books on theology are read today only by a small group of specialists, his *Apologia*, an autobiography of a troubled soul in search of personal knowledge, is 'irresistible,' as Wilde says, and 'will always be dear to the world.' As fallen leaves that keep their green, the relics of a spiritual autobiography are to secular learning what the life of Moses, Christ or Mohammed is to religion. For in Wilde's words, 'whenever men see the yellow snapdragon blossoming on the wall of Trinity they will think of that gracious undergraduate who saw in the flower's sure recurrence a prophecy that he would abide for ever with the Benign Mother of his days – a prophecy that Faith, in her wisdom or folly, suffered not to be fulfilled' (1968, 198). We tend to think of education as the imparting of information in lectures and seminars. But it would be more accurate and rewarding to say that personal knowledge – including the conversations we have with imaginative instructors and the discoveries we make in archives and laboratories – *is* the education; the formal teaching is concerned with the preparation for these conversations or discoveries and with means of expanding them.

For me, personally, a bespectacled Jewish Quaker who looked like Spinoza and who lectured on that God-intoxicated Euclid in a tiny lecture room at the top of a stone tower can still touch the heart. My memory is also stirred by a rehearsal forty-five years ago of the Hart House choir, overheard from the Gallery Dining Room, and by a bemused mentor prodding me from behind clouds of tobacco smoke in a gabled garret above a row of cloisters. Even a roaring fire in a huge stone hearth beside a tree rising to the timbered ceiling of the Great Hall in Hart House remains an emblem of the lifetime I have spent at the 'great good place' which first introduced me to the life of the mind and which was to become many years later my intellectual home.

8

From Maps to Models:
Closed and Open Knowledge

Mortimer R. Kadish, a recent defender of cultural laissez-faire, advocates free trade in ideas and equal concern and opportunity for each scholar or scientist with the desire and talent to compete. But a democracy's genuine concern for the individual is always in danger of collapsing back into passive inertia and conditioning. As a cure for the drab uniformity diagnosed by Alexis de Tocqueville as the worst disease of American democracy, Allan Bloom prescribes an elitist classical curriculum (sicklied over with the pale thought of caste). And even in deriding Matthew Arnold for thinking of culture as a rearguard religious activity, T.S. Eliot supports his attack, not only upon a utilitarian theory of education that promotes the greatest happiness of the greatest number, but also upon the whole tendency of the utilitarian mind to sacrifice all other sides of our character to the business or professional side.

Though an affluent, market-driven society needs higher learning to sustain it, its universities must also oppose the intellectual conformity and lack of daring that Tocqueville identifies as Western democracy's greatest weakness. Whereas pre-college education subordinates the individual to the group, higher learning develops the insight of each participant. In a liberal democracy the only authority worthy of the name is the 'free authority' of an educated society honouring each member's special contribution. Scholars and scientists need to think of their academic calling, less as an exercise in cultural salvation (like Allan Bloom and T.S. Eliot) and less as a defence of social democracy (like J.S. Mill and Mortimer Kadish), and more as a recovery of the lost models of scholarship and the neglected paradigms of science – the way Protestant reformers used to think of the church.

Each scholar or scientist is subject to a higher authority; namely, the evolving model of his discipline. Like a believer in his apprehension of the Logos, which must be understood to be loved, every scholar is required to take part in a conversation. But this conversation is not primarily with the institutional life of a profession revolving around grading, promotions, and committee meetings. Nor is it with the intellectual mediocrity of a society whose economic growth and material well-being are more valued than its scholarship and culture. A professor's most sustained conversation is with an academic model or paradigm. Each discipline tries to rediscover and keep alive this model just as Protestant reformers try to renew the authority of God's Word in Scripture.

St Paul believes the truth will set us free. But the freedom he has in mind is not the freedom to join a sect but the freedom to be liberated by the higher law of the Gospels. As scholars we should not be too loyal to any group of Marxists, feminists, or cultural materialists who want to control what we do. The only loyalty we owe is to the freedom to test, revise, and even reinvent the models of our disciplines. As one observer says, 'group loyalty, including class commitment, is admirable in certain settings, but it has a long history of limiting intellectual freedom' (Fleishman, 90).

According to Coleridge in *Aids to Reflection*, St Paul's metaphor of substitution and atonement tells us less about Christ's sacrifice than about Paul's interpretation of it. By contrast, the analogy between members of a human family and the persons of the Trinity tells us more about the Christian God than about the theologians who interpret him. When a school of medicine or law bases its authority on professional organization and on the licensing of members rather than on scientific principles or rules of jurisprudence that regulate its daily practice, it is in danger of mistaking a mere metaphor of authority for a genuine analogy. The resulting worship of machinery is likely to be totalitarian and idolatrous. When, on the other hand, a discipline like philosophy or physics bases its authority, not on professional organization, but on models of mind or nature that it is constantly trying to reinvent or refine, it may be accused of replacing a stable analogy with a mere metaphor to be changed at its convenience or will. Critics of Richard Rorty's philosophy and of Thomas Kuhn's contribution to the history of science have attacked these thinkers for making the paradigms of their disciplines too unstable and therefore less authoritative than they truly are.

In his important book *Toward an Ethic of Higher Learning*, Mortimer Kadish argues that to be genuinely liberal, education must turn the information provided by a map into the knowledge conveyed by a model, which has both heuristic and predictive properties. The maps that are studied in a survey course chart the terrain and general contours of a subject. By contrast, a model initiates its users into the genres of discourse which scientists, philosophers, historians, or literary scholars habitually use and sometimes transform in the practice of their disciplines. Each genre of discourse has its own unique style, themes, and qualifications for participating in the game (or what Kadish – more colorfully – calls the 'dance') of higher learning.

For Kadish the 'non-vocational heart' of university education lies somewhere between 'the nightmare of the purely instrumental mind' and a 'non-instrumental use of reason for its own sake' (13). The grace or awkwardness, the ease or rigour with which a physicist uses the scientific method to solve or explore a problem in quantum mechanics is what Kadish means by the 'style' of his research. Just as important as style is the search within each genre of discourse for an organizing 'theme.' As Kadish explains, 'the French Revolution is not as such a theme but a reference ... The theme – depending on how one regards the French Revolution – might be the development of democracy or the perspectives of the totalitarian state' (57). The search for appropriate themes is a search for the permanently important concerns that seem to 'emerge' spontaneously in each branch of learning whenever 'the possibilities of human intelligence and sensitivity are pressed' (57).

Since most scholars and scientists have only a finite fund of talents, one function of a survey course is to reduce the risks of professional catastrophe. 'To succeed in grasping a style appropriate to the development of some theme, may sometimes ... ruin the person for another style, another theme' (Kadish, 60). A would-be philosopher may be too gregarious to accept the loneliness of a contemplative vocation. Conversely, a research scientist who values solitude may function ineffectively as a member of a team to build a nuclear bomb, which requires the interaction of specialists. Even an independent studies course may show the aspiring chemist that the research scientist and the monk are first cousins. Sometimes in a biologist like Gregor Mendel, the founder of genetics and the abbot of an Augustinian abbey at Brunn, the talents of the scientist and monk converge in a single person. A neurologist who spends her life seeking a cure for multiple sclerosis or Parkinson's disease may have a volatile temperament or an impatience with delay that unfits her for the postponed outcomes and setbacks that a scientist must learn to live with. As Kadish explains, the maps of a survey course may provide 'an initial insurance against ineffectiveness and incoherence in both education and life, and therefore some provisional insurance against certain palpable kinds of injury to the interest in self' (69).

The greatest challenge in higher learning is not a lack of information but an inability to handle models that require contradictory talents or incompatible skills. The merely probable reasoning of an orator would be a defect in a mathematician who is required to give strict logical demonstrations. Conversely, strict proofs may be less important for a defence attorney than an imaginative capacity to enter the mind of an unsympathetic juror or unfriendly judge. Above all, a talented scholar or scientist must be prepared 'to endure risk and uncertainty while taking pleasure in the process that creates them.' Unfortunately, as Kadish shrewdly notes, 'not all temperaments can take that' (63). To be well qualified for the game of higher learning, a scholar may sometimes have to be a living oxymoron, a thinker who is well informed but visionary, cautious but ambitious, scrupulous yet inspired. Perhaps only prophets, rebels, and poets, scholars who are simultaneously 'stubborn and humble,' 'contentious and open,' or 'self-contradicted' (64), as Kadish says, are fully equipped to play the game.

Unlike a scholarly model or a scientific paradigm, a map has nothing to say about final causes or goals. Though a map may prevent a scholar from spending 'a lifetime walking in the wrong direction,' an education wholly devoted to map-reading or survey courses could easily entangle him in 'a network of routes.' Such learning is tantamount to entering a library and, instead of reading any literature or science, spending all one's time devouring an encyclopedic 'catalogue of catalogues' (O'Brien, 35). Maps and catalogues are non-directive: if we know where we want to go, they can tell us how to get there. But they provide no habits of mind and no motives. In Kadish's words, they can 'locate possible places to visit or reside in, the relation of these places to one another, and how one might get to those places' (69). But since a map-user as such never participates in the activity of map-making, his life may be full of destinations and still lack a destiny. By contrast, a model allows its users to conserve, modify, or even transform the material they study. Because a model may even change a scholar's habits and incentives, it may give his life a goal.

Sooner or later learners feel impelled to turn toward models, which might be defined as maps devised specifically for the purpose of their users. One such purpose might be to provide the learner with an analysis and evaluation of his culture. It may encourage the intellectual historian or philosopher to ask if some cultural values worthy of being pursued as ends in themselves – values such as equality and liberty, justice and mercy – necessarily exist in inverse ratio. Is it true that the more we have of one, the less we have of the other? It may even prompt historians to ask if such social blueprints as Plato's *Republic* or Matthew Arnold's *Culture and Anarchy* can reconcile the virtues of the moral and intellectual life. Or is every choice between Christ and Plato, Jerusalem and Athens, likely to entail some irreparable loss?

Most scholars have little patience with geniuses like Browning's Paracelsus, a Prometheus too clever in general to master any map or model in particular. Scholars are more likely to sympathize with Browning's unglamorous but dedicated grammarian, who accepts a Renaissance humanism of patience and delay. Since many cherished projects fall into oblivion, we tend to identify with a scholar who pursues a goal he never attains. If the scholar's bodily decrepitude is not exactly wisdom, it signals at least a grim determination to grind away at grammar even as he utters his death rattle. But as language winds down and the hope of producing anything more enduring than a footnote to a Greek or Latin grammar book begins to die, there is more grotesque irony than praise in Browning's portrait. Nothing can quite close the gap between the plodding pedant who, 'adding one to one,' reaches a hundred, and the soaring genius, a Paracelsus in disguise, who, 'aiming at a million,' fails to score a single point ('A Grammarian's Funeral,' ll. 117–20).

Without a model, Francis Bacon warns, an apprentice learner is in danger of losing his way in 'a kind of wandering inquiry, without any regular system of operations' (432–3). Even when the 'throttling hands of death' clutch at his throat, Browning's grammarian can produce only delicate or effeminate learning. He continues to grind away at 'the doctrine of the enclitic *De*, / Dead from the waist down' ('A Grammarian's Funeral,' ll. 131–2). To be an innovative linguist, Browning's pedant must learn to view his subject from the vantage point of some theory or model. He needs to climb an observation tower from which a mere attentiveness to syntax and grammar can begin to assume the contours of a theory of language, living and supple. Only from such a platform can a scholar like Christopher Ricks show in *The Force of Poetry* how self-reflexive tropes are the chosen idiom of a poetry of insurrection and civil war, and how anti-puns fence off violent or brutal meanings in the poetry of Robert Lowell. The steps of a true scholar-critic 'must be guided by a clue,' as Bacon says, 'and the whole way from the very first perception of the senses must be laid out upon a sure' model or 'plan' (434).

In addition to providing a useful resting-place and goal, a successful model must be responsive to change. Though fashion may be the first refuge of ignorance, a discipline like literary criticism may have to tolerate a rotating circus of Marxist acrobats, feminist tightrope walkers, and structuralist clowns if it hopes to survive its fads and renew its legacy. As Kadish observes, 'the possibilities of reciprocal engagement and happy competition are prime factors in considering what subject matters are to be chosen' (82). But 'a maximum of appropriate alternatives' also carries an irreducible minimum of risk. Charismatic feminist critics and ideological gurus may promote new scholarly models that their acolytes too uncritically embrace. The scholar as exhibitionist draws attention to herself rather than to the model she should test or explore. 'Educational supersalesmen are not to the interest of either student or instructor,' Kadish warns. As Pygmalion, Professor Higgins makes a poor instructor, since Eliza Doolittle should be 'nobody's Galatea, no matter how benign and talented the artist' (82).

Other criteria for selecting models are their ability to be 'complex, subtle, and demanding' (Kadish, 83). One reason why Petrarchan and English sonnets are models of discourse worth mastering is that in the sestet of each sonnet such masters of the form as Shakespeare and Milton are able to resolve or transcend the problems they explore in the octave. Over the centuries the English sonnet conventions have been disturbed or transformed, often in response to the shock of cultural change. Since many subtle, complex poems have been written in this form, the model itself has proved remarkably fertile and enduring.

To become fully intelligible, a model has to be tested and applied. Though the conventions of a literary genre can be summarized and committed to memory in five minutes, it may take a lifetime of study to understand how a set of norms is native to a group of great writers. I have spent a decade of my life writing essays and books on the origins of elegies and dramatic monologues. I have also offered seminars that supplement the maps of an impoverished literary history with an understanding of how mask lyrics, dramatic monologues, and monodramas can provide a better understanding of the bad faith, casuistry, and mysteries of unconscious motivation that allow some of the most inventive poets of the last two centuries to reinvent a protean genre. I am equally engaged by the conventions of pastoral elegy, elegiac lyrics and confessions, because in exploring life-and-death issues from one generation to the next they show how strong and weak mourners fluctuate between subdued elation and wild, helpless grief. Long before Freud, elegies reveal how the work of mourning, in literature as in life, can even illuminate what one commentator, Reynolds Price, calls 'the nadir of mystery,' the 'killing stretches of our lives in which God is silent and, in that silence, appears to torment us or someone near to us for no reason discernible by the human mind' (52).

I am haunted by genres, possibly because any model worth mastering depends upon deep study and discipline that are simultaneously verified or 'proved' upon our pulses, like a convention in poetry or music. As an example of what Frye calls a 'rhythm of attachment' and a means of 'stylizing life,' a genre like pastoral elegy can detach both Milton and the reader from a painful experience of premature death in 'Lycidas.' As Frye says, 'Bach, Handel and Mozart can tear you to pieces emotionally by just following the rhythm of a convention' (36). Unfortunately, the two-dimensional maps of a textbook or survey course are all that many students want or ever know. To offer a more challenging three-dimensional model to students who seek only a ready passport through the next examination is to court disaster in many student evaluations, as professors often learn to their dismay.

One problem with 'Anti-Calendars' that turn course evaluations into popularity contests among professors is that they tend to discourage a timid student from taking a demanding course on the great models of his discipline. A 60 per cent 'retake' rating in a course designed to stretch the mind instead of drugging it with cribs or burying it under a mound of mind-numbing maps too often sends to students the wrong kind of signal. For even in introductory courses, a resourceful use of laboratories and tutorials can provide the more ambitious and committed students with the rudimentary contours of a country still to be discovered and explored, instead of a flat and distorting Mercator projection of familiar territory.

Paradoxically, the more perfectly a model has been assimilated and mastered, the less transparent it will be. If I am looking for a model of pastoral poetry, I do not turn to a great pastoral elegy like 'Lycidas,' which is *sui generis*, but to a minor eclogue by Theocritus. We might risk a paradox by saying that it is precisely because Tennyson's vision in *Maud* is so authentically Spenserian and pastoral that it ceases to be consciously modelled on Spenser. As Tennyson, like Spenser in *Epithalamion*, achieves the symbolic stature of a miniature biblical epic in which Maud is both Eve and Mary, the virgin of the *hortus conclusus* of the Song of Songs, we sense that these garden lyrics are less the imitation of a Spenserian model than the kind of vision that all pastoral poetry – from the Bible to Spenser – exists to express.

Like religion or a joke, each model is meant to be appropriated by a particular culture at a specific time. To say that a historical romance is like a bustle because both are fictitious tales based upon stern reality is to make a pun that is possible only in English and only after the invention of a specific fictional genre and style of dress. To attempt the joke in German would be to play the verbal game without the model of the double pun that makes it feasible. When Samuel Johnson, who leaves few turns unstoned, complains that a pun for Shakespeare was the fatal Cleopatra for which he lost a world and was content to lose it, he is saying that the high seriousness of a tragedy and the low humour of a pun are incongruous models. Even in experiencing the defeat of his high hope and promise, Milton's Adam in *Paradise Lost* is tutored in the mystery of a kingdom that will make him happier far than the kingdom he has lost. Whether he is telling a joke or writing a biblical epic, a storyteller is making use of a cultural model. Without the theological paradox of a happy fall it would be difficult for Milton to acquire and organize the information of his epic. But just as Milton's God never forces Adam to fall even though he knows he *will* fall, so there seems to be a curious indeterminacy in the power of a model to shape its materials. It imposes a pattern without determining an outcome.

According to Kadish, the very indeterminacy of models is 'part of the process through which people function most profoundly in their fields' (92). Instead of providing an unambiguous map of the moral and intellectual virtues, for example, Aristotle's *Nicomachean Ethics* offers the ambiguity of a model which forces readers to make their own decisions about how to interpret it. Like most models, the *Nicomachean Ethics* is 'tougher and rawer, more challenging' than the *Eudemian Ethics*, 'its tensions and contradictions more exposed' (Sparshott, 1994, 4–5). A slavish adherence to maps encourages a pedant to do by halves or quarters what a bolder thinker like Aristotle will accomplish at a single stroke by introducing a unified theory of the soul. Instead of blunting our minds, genuine models will tease, ravish, and transform us. But a final mastery of a model like Aristotle's may elude us in the end, if only because his use of such key terms as 'activity' or 'zeal' or 'seriousness' is flexible and often looks two ways at once. 'Provided the center ... holds,' Kadish believes that the 'indeterminacy of Models, rather than obstructing the movement of the liberal ballet, will provide it with a further dimension'

(92). For even if there were world enough and time, only God himself could hope to master every model. Derek Bok is being unusually honest, I think, rather than merely irresponsible when in a frank book on higher learning he claims that ignorance alone prevents his analysing the sheer diversity and indeterminacy of models (1986, 3).

Whereas a map, like a technology, is a means to some end, each model of higher learning is an end in itself. In the act of mastering or imitating a model, the mind enlarges its scope. Until George Eliot's Lydgate sees himself as a Victorian Vesalius in *Middlemarch*, the excitement of becoming a great research scientist never comes into focus for him. And not until the narrator sees Dorothea as a modern St Theresa, or the German painter Naumann sees Casaubon as a model for his portrait of Thomas Aquinas, do their worlds begin to widen. Unfortunately, living well and living effectively are often in conflict. In 'laying' herself 'off on new coordinates,' as Mortimer Kadish would say, Dorothea finds herself trapped in a maze or labyrinth. She hopes to be liberally educated as the wife of a great scholar. But in a world where 'perfection of the life and of the work' are often painfully at odds, and where 'the quickest of us walk about well wadded with stupidity' (Eliot, 144), an education in scholarly and scientific models may doom Dorothea, as surely as it dooms Lydgate, to personal catastrophe. As Kadish says, 'the liberal education' of such people 'entails opening the self, *through* its accomplishments, to frustrations and boredom when the person lives effectively in the world' (95). But who wants to live effectively if the price of such prosperity is Lydgate's abandonment of the medical research that makes his life purposeful? Or what wife would want wealth or social standing if it means marrying an affluent clergyman who is doomed to fail at the scholarly vocation that attracts her in the first place?

In her parable of the pier-glass, George Eliot compares a cultural model to a candle that has power to turn the scratches of random events into 'the flattering illusion of a concentric arrangement' (195). But the parable never allows its interpreters to forget that their mastery of a model may banish incoherence at too high a cost. For there is always a multitude of other candles that the selection of any single centre of illumination forces the observer to exclude. Since the illusion of order is only a reflection of the observer's ego, each model may tell us more about the interpreter than about the world she interprets. In Kadish's sombre words, 'all but the luckiest of those in liberal education are impaled more or less painfully upon the choice of making out or going for what they want.' Those like Dorothea or Lydgate 'who place their bets on mind might conceive that they just might win their wager against disjointedness in their lives or, losing it, provided they are lucky, gain a shot at creativity' (95). But like George Eliot's most attractive characters, the more ambitious and single-minded the competitors become, the greater their chance of turning into 'superfluous people,' into anomalies or social misfits.

It may take years before a scholar finds he is no longer impersonating a literature or law professor but using his discipline's patiently acquired models to function independently as a philosopher of law or as a literary historian. In the meantime a scholar's disillusion with his performance and even his silent envy of a rival's achievements may remind him that rank, honour, and role-playing have no importance compared to the rewards of contributing to knowledge and of thinking and living like a scholar, which is the prospect that drew him to academic life originally.

Though participation of any kind may be preferable to silence, anyone who tries to take part in common room conversation knows it is easier to drift into gossip about colleagues than sustain lively exchanges about Merleau-Ponty or Heidegger. I prefer to talk at the student cafeteria about Pessoa's heteronyms and Browning's masks than discuss grade inflation at the high table. Seated between Northrop Frye and Francis Sparshott in Burwash Hall, I was often reduced to a stammering one-liner: 'Pass the salt.' My comment was hardly a contribution to the 'conversation of mankind.' Since communication across gaps of intellectual endowment and knowledge is at once necessary and impossible, all scholars dream of an ideal conversation. I remember with affection the two-hour conversations I used to have after lunch with a small band of colleagues and friends in the Senior Common Room. One friend said they reminded him of conversations in a Russian novel. I also remember a seminar I gave on Friday afternoons that used to reorganize itself spontaneously into a continuing conversation in a local coffee shop. Even so eminent a luminary as Albert Einstein said that he went to the Institute of Advanced Study at Princeton mainly for the pleasure of walking home with Kurt Gödel. But why is the substance of searching conversation – during walks and seminars, in common rooms and college dormitories – so seldom reproduced in modern biographies? Is it because there are fewer such exchanges to record than in the age of Johnson or Coleridge? Or is it because when memorable conversations do take place, there are fewer James Boswells or Michael Ignatieffs to record them?

By 'explanatory discourse' Michael Oakeshott means any model of disciplinary thought that informs and gives shape to an intellectual inquiry or to an experiment in science or living conducted under its auspices. The discourse of philosophers, for example, stands in the same relation to both the empirical project of Locke and the rationalist project of Spinoza as the Spanish language stands to a literary master-piece like *Don Quixote*. Though keeping alive an explanatory discourse is the proper function of a professor's scholarship, this function will usually differ significantly from the proper aim of university teaching, which is to instruct students as clearly as possible in the 'paradigms' or models of a discipline. Only rarely, at advanced levels of teaching, can professors follow Gerald Graff's advice to 'teach the conflicts' that arise in the explanatory discourse itself. Oakeshott is right, I think, to argue that the paradigms and their conflicts must be lived or embodied rather than talked about or analysed.

In *Professing Literature* Graff seems to think that studying the history of how literature has been taught is identical with studying the maps and models that have shaped the growth of literature itself. But pedagogy is not literature; and what is learned in a history of English departments is not identical with a theory of literature. A study of the former may dis-close a cycle of revolutions and renewals: a mindless empiricism, for example, may prompt a call for theory, which fossilizes in turn into a new empiricism. As Graff himself points out, 'defenders of theory tend to equate the New Criticism itself with unreflective empiricism, but in its time the movement stood for theoretical reflection against the primitive accumulation of data' (247). The effect of these cycles, however, is not to consolidate the authority of 'institutional history,' which is the subtitle of Graff's book, but to send the scholar in search of better, more enduring paradigms. Graff's error is to mistake the history of pedagogy for a his-tory of the explanatory discourse literary scholars are always trying to master. The net result of Graff's project is to move the faculty of educa-tion from its proper place at the edge of the university to its centre.

I am not denying that scholars may find it useful to recognize and per-haps even alter the repetitive cycles in their discipline, especially if these cycles are non-progressive. But to teach the conflicts of pedagogy is to teach little more than academic politics. And who wants to wash depart-ment laundry in public? Perhaps pedagogical conflicts may profitably be dredged up and studied in a seminar on teaching. But such a seminar is not a history of literature or science as such. Indeed the most important models in a discipline are the most difficult to theorize because they are the most elusive and fugitive. They are the hardest to stabilize and the trickiest to identify or analyse accurately. To become self-conscious about models and institutional practices is often to misrepresent their nature and falsify their value.

Lines of inheritance must be kept intact, not because all traditions should be preserved, but because some of a culture's capital should be reinvested in keeping alive and renewing the conversation of mankind. The university is always more than an archive or museum. But the human mind (though rich in its accidental varieties) is poor in its essential types. Aristotle refines a universe of discourse that Aquinas applies to theology in the Middle Ages and that a modern commentator on Aristotle, Francis Sparshott, uses to criticize contemporary third-person views of the world. Indeed Nietzsche's apology for history is timely in its very explication of what is 'untimely' in the past's ever-changing relation to the present.

The philosopher Robert Nozick has suggested that an explanatory theory may promote understanding even if science later rejects the theory as untrue. Ptolemaic astronomy may be false. But it can still teach many truths about geometry and about the ingenious fine-tuning of an explanatory system designed to make sense of recalcitrant data. Nobuhiro Soda, a physics undergraduate who in weekly tutorials with me compared models in science with metaphors in poetry and myth, took pleasure in reinventing the calculus as Leibniz rather than Newton developed it. Nobu's pleasure may be untranslatable into anything less delightful or gnomic than mathematical invention for its own sake. Nevertheless, his alternative calculus may solve future problems in physics better than Newton's method. In his book *Philosophical Reasoning*, John Passmore shows how many explanatory discourses that are not scientific or 'provable' are still philosophically meaningful. No academic discipline is more conscious of its search for models than philosophy. Whenever I used to serve as an outside examiner in the philosophy department, some scholar would say, 'That is an intriguing line of inquiry, but is it philosophy?' Perhaps books like Passmore's and Nozick's will keep open the question and encourage logicians, moral philosophers, and metaphysicians (if they still exist) to write more essential, if not better, philosophy.

Helen Vendler's introduction to *The Music of What Happens* and her essay 'The Function of Criticism' (1988b, 16) advance literary knowledge by developing an aesthetic criticism devoted less to philosophy or politics than to the contemplative pleasure afforded by other arts like music and painting. And in 'The Three Voices of Poetry' T.S. Eliot injects new life into a study of genres by unexpectedly contrasting lyric and drama, not with epic (as in most traditional discussions), but with a new rhetorical genre that is more dramatic than a lyric poem but more lyrical than a stage play. Eliot tries to bring this genre to life in the choruses of his play *Murder in the Cathedral* and later (with more success) in his Ariel poems, 'The Love Song of J. Alfred Prufrock,' and his dramatic monologue 'Gerontion.'

Unlike the paradigm of a first-conjugation verb in Latin, which is Thomas Kuhn's example of a pure paradigm, a scientific paradigm like Newton's laws of motion is not really a paradigm (in the strict grammatical sense) but an extremely stable model. The fact that models change more rapidly in scholarship than in science may reflect a measure of chaos in the former. Or it may be a warning that scientific paradigms seldom apply to scholarship. The aspiring theologian or philosopher must usually commit himself to the models and protocols of his discipline – and acquire a 'feel' for them, as we say – before he can tell where exactly they will take him. In a phrase favoured by Augustine and Anselm, the scholar has to believe or have faith in his model before he can understand it: 'credo ut intelligam.'

Though historians of science like Lewis Thomas and Stephen Jay Gould write splendid monographs, a full-length book would damage the reputation of most scientists, whose refinement of a stable paradigm can be better accomplished in a concise, densely researched article for a scientific journal. If all scholarship aspired to the condition of a science, the university presses would soon go out of business. In disciplines where books rather than learned articles are the preferred mode of publication, the rapid fluctuation of models may confirm the priority of *proposals* to *proofs* and may even suggest, in Richard Rorty's words, a 'lack of international consensus about who is doing worthwhile work' (1999, 181). David Perkins's chronicle of the ever-changing methods of writing literary history, for example, contrasts sharply with Thomas Kuhn's chronicle of the remarkable stability of most scientific paradigms. Models that are too unstable may produce the vertigo of literary studies during the 1990s. But a model that is too stable may produce only the predictable hack work of a technician.

In famous passages in the *Ethics*, the *Metaphysics*, and the *Parts of Animals*, Aristotle declares 'that the ability to judge what degree of precision may fairly be expected in any inquiry is the mark of an educated man: such a person would not accept probable reason from a mathematician or require demonstration from an orator' (quoted by Culler, 203). Applying Aristotle's dictum to modern paradigms, we might say that the accurate measurements required by an ophthalmologist who is about to perform laser surgery for glaucoma are inappropriate to jury members who have to determine beyond the far more flexible limits of a 'reasonable doubt' the guilt or innocence of a woman accused of murder. Isaiah Berlin is speaking as an Aristotelian when he observes that 'degrees and kinds of precision depend on the context, the field, the subject-matter; and the rules and methods of algebra lead to absurdities if applied to the art of, say, the novel, which has its own appallingly exacting standards' (167). To learn to write prose fiction, literary criticism, history, or philosophy is to master distinct models of discourse without making informed (or invidious) judgments about the relative value or difficulty of the different genres. As Berlin observes, 'the precise disciplines of Racine or Proust require as great a degree of genius, and are as creditable to the intellect (as well as to the imagination) of the human race, as those of Newton or Darwin or Hilbert, but these kinds of method (and there is no theoretical limit to their number) are not interchangeable' (167).

To help the learner of philosophic habit avoid two extremes – both 'reason run mad' (or the mere viewiness which builds castles in the air) and 'memory forgetful of its place' – a genuine model should provide a 'judicious admixture of reason and memory, of fact and system' (Culler, 197). To master a model that provides a 'view,' it is not enough for a scholar to acquire knowledge passively by swallowing wafers or mixing letters, as Swift's scholars do in the Academy of Lagado. Nor is it possible to dispense with the arduous labour of amassing data. Since models often require their masters to conduct an experiment or to judge a generalization by checking records in the archives, a scholar's achievement seldom justifies the gibe that 'those who can do, those who can't teach.' By placing 'us in that state of intellect,' as Newman says, in which we can 'take up' any number of subjects 'with ease and readiness' (quoted by Culler, 203), a model may even allow us to do in principle or theory what we have not yet tried to do in practice. A model is also a heuristic tool that allows us to acquire knowledge we do not yet possess.

Dwight Culler has argued that the three great mental crises in Newman's life are partly a result of Newman's failure to move successfully from maps to models. In his portrait of the man of philosophic habit, Newman illogically defines perfect knowledge as a 'clear, calm, accurate vision and comprehension of all things' (Culler, 191), as if it were possible to know every subject thoroughly. To avoid being too narrow, the scholar must shun the unassimilated facts of a survey course or map. And to avoid the opposite error of being superficial, which is the danger Newman's mentor Richard Whately runs, he must beware of the empty generality of an untested model or theory. Too many scholars who look for a 'key to all the keys' lose themselves in a maze of pen scratches like George Eliot's Casaubon. They are devoured by what they devour: the labyrinth eats them up.

Ideally, Newman's scholar is a citizen of the world, a person who, in Jeffrey Hart's phrase, is capable of 're-creating his civilization,' not because he is a Renaissance man, but because he understands 'his civilization in the large, its shape and texture, its narrative and its major themes, its important areas of thought' (ix, x). Unfortunately, this paragon is being asked to do what is not strictly possible: in an age of specialists, he is asked to 'see life steadily and see it whole' (Matthew Arnold, 'To a Friend,' l. 12). In ordering his mind to perform contradictory tasks, he finds it cracking under the strain. 'Thrown into a panic by some approaching examination,' as Culler says, Newman fears 'that he had not mastered his materials at all' and that his mind is 'a labyrinth' rather than a well-organized archive. Newman's disordered knowledge is 'symbolized by the very form his illness took – he felt "a twisting of the brain, of the eyes," and it seemed that his "head inside was made up of parts" – as if the divine vengeance were chastising his arrogance and confirming him in the very state in which he had sinned' (Culler, 205). The more facts Newman tries to commit to memory as a walking encyclopedia, the less steadily he can command a view of the whole circle of knowledge. Perhaps Newman should have been more tolerant of J.S. Mill's 'one-eyed men,' especially if 'their one eye is a penetrating one' (Mill, 1950, 65). If Newman had tried to see the world more selectively, he might have gained in stability what he lost in scope. He might have sacrificed ambition to consistency. Instead, his use of two contradictory sets of images – the receptacle made up of atomic parts and the squeezed or twisted organism – shows how difficult he finds the task of substituting the models of a discipline for a mere map-like overview of its subject matter.

In one set of images the mind is merely a data bank or 'storehouse': the bigger its retrieval capacity the better. In the other set, the mind is a living organism, either a stomach that digests food or a plant that can be grown. A problem arises when Newman tries to combine the two discrepant images. When viewed as a data bank that stores memories accurately and inclusively, the mind must be stretched out and enlarged like a rubber container or elastic band. But as a living organism with a single unifying function, the mind is misshapen and eventually torn apart by the strain of being stretched and pulled open to hold as much information as possible. In Newman's own words, 'we talk of [different tasks] *distracting* the mind; and its effect upon me is indeed a tearing or *ripping open* of the coats of the brain and the vessels of the heart' (Culler, 206). It is no wonder that the challenge of becoming a Victorian Sophocles, who sees life steadily and sees it whole, should produce what Culler calls 'the five shattering illnesses' of Newman's 'undergraduate and liberal years' (205).

Constantly in search of a discipline's informing model, Newman fears that without the symmetry and design of some superintending 'view,' mere knowledge or information 'as it lies in the rude block' will never be systematic or organized enough to form the philosophical habits of a scholar (quoted by Culler, 194). Though a model's view must be broad and inclusive, it must also be accurate. A mere map abstracts too much from a terrain's variegated shapes and contours. Whereas a map provides system without detail, mere learning gives detail without system. Only a unifying model can combine accurate vision with inclusive detail. In practice, a knowledge of method consists of three parts: a mastery of models or first principles; a knowledge of how much precision is required by a model (the probable reasoning that is acceptable in oratory, for example, is inappropriate in mathematics); and a mastery of logic.

But the more Newman searches for the elusive soul of higher learning, the more it eludes him. Ideally, Newman's scholar should know everything. But in practice it is impossible for anyone to embody in himself the whole circle of learning, which only the university at large can embrace. As Dwight Culler wittily observes, 'Milton's scheme of education was not a bow for every man to shoot in, but Newman's is not a bow for any man to shoot in, and to say that it is an ideal is not to justify it' (191). Indeed Newman's famous portrait of the gentleman scholar who 'has his eyes on all his company,' who 'is tender towards the bashful, gentle towards the distant, and merciful towards the absurd' (Culler, 239), is a subtle exercise in irony. The portrait is not the solemn encomium it is often taken to be, but a playful contradiction in terms, a kind of private joke. As a *reductio ad absurdum* of negative capability, Newman's labile chameleon can be 'all things to all men,' as Culler concludes, 'because he is really nothing in himself' (239).

Huxley's portrait of the educated scientist, a cross between Newton and Shakespeare, an impassioned Keats and a sober Milton, is just as utopian as Newman's portrait of the scholar. The mind of Huxley's scientific paragon is 'stored with a knowledge of the great and fundamental truths of Nature and of the laws of her operation.' Though he is 'full of life and fire' and 'no stunted ascetic,' his passions are also 'trained to come to heel by a vigorous will, the servant of a tender conscience' (3:86). It is clear in such a passage that Huxley the scientist, 'whose intellect is a clear cold logic engine' (3:86), is seduced by the metaphors of Huxley the rhetorician. If his mind is best described by a technological model, 'ready, like a steam engine, to be turned to any kind of work' (3:86), we should not be surprised if the work to which it turns is scientific, since scientific technology furnishes the language of the mind's description. The problem is not merely that Huxley, like Newman, combines discrepant pictures of a mind that must be simultaneously a spinner of 'gossamers' and a forger of 'anchors.' The more serious problem is that Huxley assumes in his major premise what he should be proving only in his conclusion. He argues that a scientific education trains the mind because, as a rudimentary computer, the mind is already scientifically inclined. One can see what George Eliot means when she says that, if Aristotle had been 'the freshest modern' instead of 'the greatest ancient,' he would 'have mingled [his] praise of metaphorical speech, as a sign of high intelligence, with a lamentation that intelligence so rarely shows itself in speech without metaphor' (1961, 124).

To resolve the problem that baffles Newman it may be necessary to refine our earlier analysis of maps and models. A map is a model only in the trivial sense that, once we know how to interpret one Mercator projection, we know how to interpret all the others. But a map of North America has no predictive power: unlike a model of DNA or a literary genre like the sonnet, it tells us nothing about the shape or contour of a different continent – of a *terra incognita* like Antarctica, for example. In Kantian terms, the logical structure of a map is *synthetic a posteriori*: it is genuinely cognitive, because its predicate provides information not contained in the subject. It has no predictive or heuristic power, however, because its knowledge depends wholly on the experience of our senses: a new continent requires a new map.

A model, by contrast, furnishes essential ingredients of what we can know. Most of the body's one hundred trillion cells, for example, contain a nucleus with forty-six chromosomes (long, coiled up strands of the molecule DNA), twenty-three from each parent. The model tells us *a priori*, in advance of studying the cells of any individual person, that a typical strand of human DNA contains more than three billion chemicals called base pairs, which make up the digital letters in the genetic code. What cannot be known prior to experimental testing, however, are the precise three-letter 'words' of the messages that only a small fraction of the chemical base pairs will transmit. Nor is there any way of knowing *a priori* which parent's chromosome will provide the active coded instruction to build hair texture or brain and kidney tissue. In Kantian terminology, the human genome, like any model, has a *synthetic a priori* structure. Because each human being is unique, every code of life will add new biological information in the predicate that its subject does not contain. But just as we can predict in advance that a sentence written in English will consist of words defined in the *OED*, so we can identify *a priori* the digital letters of the chemical base pairs in which each genetic code will be written.

One of the most vicious distempers of higher learning is the belief that a map or model is the cultural equivalent of Richard Dawkins's 'selfish gene,' which behaves like a virus in digital technology. Like a computer code, which consists of a string of ones and zeroes, a gene is digital because its DNA is a string of chemical letters. But as it passes through different minds, a cultural model is too fluid and protean to transmit its code by means of chemical letters or numbers. As a scholar who has worked for three decades with literary genres, I have still to find a cultural model that reproduces itself like a selfish gene. Since I admire the work of T.S. Eliot and Robert Langbaum, I once tried to imitate their theories by writing essays that resemble 'The Three Voices of Poetry' and *The Poetry of Experience*. But because I tested different dramatic monologues, the end results were disappointingly different. According to Dawkins, a cultural model is most likely to behave like a replicating virus or a selfish gene if the copyist follows a set of instructions telling her how to fold origami paper instead of trying to copy the product, the finished crane. But not even the most slavish devotion to a blueprint or a map can achieve the mindless reproductions of digital technology. If I am looking for the cultural equivalent of a selfish gene, or what Dawkins calls a 'meme,' I find it, not in the archives of higher learning, but in the infamous annals of bigotry and prejudice, where it blindly propagates itself as a destructive virus of the mind. It seems to me that any scholar who seeks a true and lawful marriage of higher learning's maps and models to digital technology, 'in love and holy passion,' as Wordsworth says, will soon be suing for a divorce.

If a thinker like Newman has difficulty combining the generalizing predictive power of a model of the cell's nucleus with the inclusive descriptive capacity of a map of the human genome, it is because he wants to do opposite things simultaneously. Had I been Newman's tutor, I would have discouraged him from crossing boundaries and trying to make all learning his province. At the risk of displaying the presumption of a minor sophist lecturing Socrates, I would have advised Newman to start instead with models of a particular discipline like theology, ethics or human biology. He could then have used those models to make his own moral and educational discourse as inclusive and heuristic as possible.

All education depends on two equally important tendencies – one restraining, the other encouraging, the criticism of models. If scepticism has no play, the established models of a discipline like history or physics will cease to be tested or defined, and learning will cease to advance in these areas. But if the all-dissolving, all-corroding scepticism of the intellect breaks out of bounds, as Newman fears it will in matters of religious inquiry, then a discipline like theology or metaphysics may perish for lack of any models to test.

A model is mastered most effectively when it is a Word made flesh, a paradigm unobtrusively informing and giving shape to learned conversation and scholarly habits. When we confuse scrutiny of a discipline's models with a mere historical account of how a discipline develops, the results are usually depressing. Frye is wrong, I think, to say that professors of literature should teach, not literature, but literary criticism. Of limited professional interest is the story of how literary history challenged philology in North American English departments. Equally provincial is the story of how the New Criticism displaced history only to be challenged in turn by such exotic usurpers as structuralism, deconstruction, new historicism, and feminism. To make the academic fashions in a discipline the explicit focus of study is to 'bureaucratize' it, to subject the discipline to the control of a currently fashionable school. Or it is to submit scholarship to the political influence of a powerful professional group like the Modern Language Association of America or English Studies in Canada.

To be initiated, by contrast, into the informing models of a discipline, a scholar or scientist will need conversation as the body needs food. By 'conversation' I am thinking not only of Michael Oakeshott's never fully

attainable ideal of the 'conversation of mankind,' in which each branch
of learning – with its commitment to discovery, contemplation, or practice – plays an important cooperative role. I am thinking also of 'the virtues of curiosity, open-mindedness and conversability' (Rorty, 1999, xxi)
associated with the pragmatism of William James, John Dewey, and Richard Rorty. Equally important are the forays into *terra incognita* found in
the best essays and common room conversations of J.H. Newman and
Isaiah Berlin. As Berlin's biographer, Michael Ignatieff, explains, Berlin's own 'thinking is indistinguishable from talking.' It is 'famous, not
only because it is quick and acute, but because it implies that thought is
a joint sortie into the unknown' (Ignatieff, 4).[1]

Like Johnson and Coleridge, Berlin combines a talent for conversation with a gift for scholarship. But the combination is rarer than we like
to assume. Francis Bacon says that 'reading makes a full man, conference a ready man, and writing an exact man' (130). It is better to be
quick-witted than slow, but it takes painstaking research into the past to
be exact. The virtues of the journalist and historian exist in inverse
ratio. A scholar must prefer solitude to company and spend a great deal
of time in the archives expanding his knowledge. A conversationalist
who would rather talk than study is likely to be a prompt and witty
speaker. But as Bacon warns, a conversationalist who has written little
should have a great memory, and if he reads little, he 'need have much
cunning, to seem to know what he doth not' (130).

Any scholar who tries to discriminate among models and paradigms
in different disciplines has to distinguish between two kinds of 'proof':
logical demonstrations and proposals to be entertained and tested.
Whereas the physicist proves (or logically demonstrates) a proposal
about the wavelike properties of light, a philosopher or a poet proves
(or puts to the test) a metaphorical proposal to express matter in terms
of spirit, for example, or spirit in terms of matter. As metaphorical proposals that invite free assent increasingly replace logical proofs that
try to coerce agreement, an academic discipline may try to 'prove' its
proposals by experimenting with them on the probation or 'proving'
ground of life itself. Such a transition from proofs to proposals supports
Richard Rorty's claim that 'the learned professions' are 'marked by an
increased willingness to admit that there is no single model for good
work in an academic discipline.' As Rorty says, 'the criteria for good
work have changed throughout the course of history, and probably will
continue to change' (1999, 181).

To say that 'the exception proves the rule' may mean, not that the exception establishes the rule's validity, but merely that it passes the rule through a crucible of lived experiment or testing. To be 'proved' true it is enough that a proposal should help solve intricate puzzles and improve the quality of rigorous or liberated thinking in a discipline. Einstein's new proposal about the nature of time in his theory of general relativity describes a time that is affected by movements in space, unlike Newton's concept of time, which is absolute. If the geometry of space and time can be changed by the behaviour of objects moving inside its four-dimensional coordinates, then, as Einstein wryly concludes, we do not *become* but simply *are*. But if the distinction between time past, present, and future is only an illusion, then 'all time is eternally present,' as T.S. Eliot says, and 'All time is unredeemable' ('Burnt Norton,' 1.4–5).

Thomas J. Kuhn's theory of paradigms is a logical corollary of Einstein's insight that time is relative and change an illusion. As Richard Rorty says, 'Kuhn suggested that in all areas we could drop the notion of "getting closer to the way things really are" or "more fully grasping the essence of ..." or "finding out how it really should be done"' (1999, 187). But Einstein, I think, is less sceptical than the Kuhn whom Rorty purports to describe. Einstein is not saying that time cannot be understood, only that it must be treated precisely like space in a geometry that includes time as its fourth dimension. Indeed Einstein would argue that his new geometry works better than Newton's because it allows him to predict what Newton cannot predict: the curvature of a beam of light in the vicinity of the sun's mass.

Without an initial trust in both the universe that exists and a hypothetical one, it is difficult to see how either universe can be known. In order to understand what happens to an astronaut's experience of time when a rocket ship hurtling away from earth approaches the speed of light, it is important to understand Einstein's special theory of relativity: as speed, mass, or strong gravitational fields increase, time slows down. But in order to grasp the full implications of this theory, it is equally important to imagine a hypothetical or counter-universe in which an astronaut travelling at the speed of light would exist in an eternal present. At 87 per cent the speed of light, the space traveller would see into the future because he would see time pass twice as quickly for the rest of the world as it does for him.

We accept a paradigm or a model as we accept a postulate of faith, which (as Paul Tillich says) 'embraces' both 'itself and the doubt about itself' (quoted by Polanyi, 280). Michael Polanyi concedes that 'a tacit doubt, an inarticulate hesitancy, like that of a marksman dubiously pulling the trigger,' accompanies every entrance into a new and strange domain of thinking, where all proposals must first be taken on faith (280). For all committed thinking describes a circle: it posits as its first principles or axioms the proposals that more linear thinking tries to reach only as the conclusion of an often protracted chain of logical demonstrations or proofs.[2] In trying to restrict learning to the few propositions that linear thinking can demonstrate, a distrust of circular thought has made it difficult to grasp the force of Augustine's saying that we cannot understand a proposition unless we first believe it. Michael Polanyi affirms that only a faith in the validity of what we know but cannot prove by means of logical demonstration can align religion itself 'with the great intellectual systems, such as mathematics, fiction and the fine arts, which are validated by becoming happy dwelling places of the human mind' (280).

The mystics define God as a circle whose centre is everywhere and whose circumference is nowhere. We can use a comparable definition to describe that circle of committed thought in which 'every field of knowledge,' as Frye says, 'can expand into other structures' (1988, 10). As soon as a student who 'takes' a subject is taken up by it, it can become 'a center of [all] knowledge' (Frye, 1988, 34). But it is hard to enter a circle of committed thinking without first cultivating the acquaintance of a gifted master or mentor. As Newman says, 'to become exact and fully furnished in any subject of teaching which is diversified and complicated, we must consult the living man and listen to his ... voice' (1856, 12–13). Why this should be is not at first apparent. Francis Sparshott compares a live lecture to a concert performance with all its attendant risks and potential distractions. But being at a concert is preferable to listening passively to a CD player, because a living master is presumably more committed to his performance than a dead one. When Harold Bloom found that Northrop Frye's books and lectures had ravished away his soul, it was not enough for him to know the wise man at a distance: he wanted to come to Toronto to study at the master's feet. Newman even believes that great books, 'the master-pieces of human genius,' are written in universities or conceived in centres of learning where scholars can come together to talk with each other in large 'assemblages and congregations of intellect' (1856, 14). Without the personal contact and apprenticeship, the Socratic inquirer may degenerate into a clever sophist or charlatan who aims 'at originality, show, and popularity, at the expense of truth' (Newman, 1856, 110).

Maps possess what John Passmore calls a closed 'know-how capacity' and models an 'open' one. All 'know-how' capacities like cloning, when divorced from wider human concerns, pose a threat. A biochemist who conducts stem cell research to cure disease may also abuse that capacity by producing human embryos in defiance of moral or legal prohibitions. As Passmore observes, the great exponents of 'know-how' capacities in ancient Greece are the sophists. Prometheus may be a hero, like Socrates. But in Faust we meet the sophist who is willing to sell his soul to the devil. And in Simon of Tournay, who boasts of using his dialectical skills to refute as well as support Jesus, we are reminded that mental agility alone is not an open capacity like wisdom, but a potentially closed form of intelligence, a kind of bigotry or religious zeal in reverse. Even Newman's cherished reason, which in its free play 'is one of the greatest of our natural gifts,' becomes in its closed corrupted form an insatiable monster. So 'aggressive, capricious, untrustworthy' is this demon that in order to 'rescue it from its own suicidal excesses' Newman has to appeal to a theological equivalent of Perseus or St George, a superintending deliverer powerful enough to save Andromeda and kill the dragon.

> I say that a power, possessed of infallibility
> in religious teaching, is happily adapted to be a
> working instrument, in the course of human affairs,
> for smiting hard and throwing back the immense energy
> of the aggressive, capricious, untrustworthy intellect.
> > Newman, 1968, 189

I am still appalled by these words from the *Apologia*. Though I possess them by memory, they never lose their power to move and unnerve me. For I know what Newman means. To develop the capricious intellectual capacity of a student prodigy is often to unleash an arrogant and unfeeling robot on the world. If a gifted student is not more than the sum of his mental powers, he is at best a monster like Frankenstein's. One way to avoid the decline of Socrates into a sophist or of Prometheus into Faust is to develop 'open' rather than 'closed' capacities. Instead of relying exclusively on textbooks, cribs, or 'maps' that reduce learning to a set of rules to be memorized and then applied by a machine, an apprentice should be encouraged to master the models of at least one major discipline. Unlike a two-dimensional map, a contoured model encourages students to honour the adventures in risk-taking that liberate the mind and keep it human.

According to Passmore, a 'closed' capacity like counting or map-reading allows total mastery of a simple skill. A grade 3 student can add and subtract as efficiently as Einstein. A scientist with a calculator may add faster but not more accurately than a novice. It is also possible to master a game with a closed capacity like Xs and Os by placing the first X in the centre square. Once the secret is known, however, no player can lose or win a game except by being careless. By contrast, no participant ever completely masters a game like chess or a discipline like philosophy or literary criticism. Though it is possible for a chess player to master certain openings and end-games and for a critic of poetry to acquire rudimentary skills in rhetorical analysis, one 'cannot "learn the secret" of writing poetry or of doing philosophy' (Passmore, 1980, 41). If a secret does exist, it is so complex that until a master poet or philosopher embodies it in his performance no apprentice is likely to intuit what it is. Alternatively, an activity like gambling may be so fraught with risk that it is difficult to improve one's performance no matter how long one plays blackjack or roulette. It is impossible to develop any capacity at all in a game of snakes and ladders for the same reason that, unless one can master chaos theory, it is difficult to get rich as a trader.

However complex it may be, a closed capacity for operating a computer or performing simple surgery can be converted over time into a routine. In general we *train* students to master a closed capacity like standard tonsillectomy operations or the use of a microscope. But we *educate* them to master a more 'open' capacity like the diagnosis of a rare disease or the mastery of a model requiring 'mathematical inventiveness,' 'historical judgment,' 'scientific imagination,' or 'literary sensitivity' (Passmore, 1980, 43). Though any closed capacity may be a necessary and sufficient condition of an open one, only a linguistic pedant would assume that a knowledge of Socrates' ideas in Plato's Greek original can be reduced without remainder to a knowledge of Greek grammar and philology. A model is better than a map for the same reason that a technique we teach our students is not so fine a possession as the wisdom Socrates embodies in his life.

The historical model that R.G. Collingwood uses to teach philosophy at Oxford challenges scholars to see texts by Plato and Kant as answers to questions that are never the same for any two philosophers. Since the answers as well as the questions keep changing with each thinker and text, Collingwood claims that his new model allows him to replace the 'putrefying corpse of historical thought, the "information" to be found in textbooks,' with an 'open' history, 'a source of unfailing, and strictly philosophical, interest and delight' (75). As soon as Collingwood replaces a closed dogmatic model with an open one, he says the rest of a tutorial hour with his students passes 'in a flash. And for myself it was no less salutary,' he maintains. 'Over and over again, I would return to a familiar passage whose meaning I thought I knew ... to find that, under this fresh scrutiny, the old interpretation melted away and some quite different meaning began to take form' (75). Collingwood finds that only the discovery of this resourceful new model allows his scholarship and teaching to draw energy from each other.

The information imparted by a competent instructor could be transmitted just as well by a literate manual or by a well-designed computer program. But to master a model is like imitating the life of Socrates or Jesus. Though the imitation of Christ may not be good enough for Emerson, Plato knows that the art of the Socratic guru or mentor is an obstetrical art that requires the apprentice to give life to something new in herself. An instructor transfers information 'from one place to another.' Or 'to use the hydraulic metaphor proposed ironically in the *Symposium*, his teaching is the syphoning of wisdom from one vessel to the next' (Gooch, 204–5). By contrast, Socrates and Jesus teach their truths by embodying them. They are difficult to imitate for two reasons. Socrates' self-effacing irony and Jesus' Messianic secret make the identity of each master a mystery. And an active response is what the mentor as midwife asks each disciple to contribute. Whereas Socrates or Jesus would be out of place in a trade school, a so-called expert or technician who tries to reduce a complex model to a mere formula or rule would learn next to nothing from Socrates or Plato, just as an unreformed pharisee has little to learn from Jesus.

Aristotle's distinction between making and doing explains an important difference between maps and models. To build a house a carpenter must acquire some skill in reading blueprints. But to produce the blueprint of Falling Waters, Frank Lloyd Wright must master the more complicated art of thinking and living like an architect. Wright's mastery of a complex model resembles the embodied wisdom of Socrates or Einstein's way of 'doing' physics: it cannot be reduced to teachable rules. Higher learning has less in common with a communicable craft like carpentry than with the architectural achievements of Frank Lloyd Wright.

Sophistry might be defined as the reduction of doing to making. It confuses wisdom with technique and a scholarly habit or way of life with rules. The opposite of sophistry is intellectual elitism, which pretends that a teachable practical science has few law-like properties and is more occult than it is. One of the worst errors we can make about Socrates is to identify him, as George Grote does, with his ideological opposite, the sophist, who replaces Socrates' devotion to contemplation and inquiry with the merchandising or hawking of a marketable commodity. To accept such a parody of Socrates as our educational model is to destroy the non-vocational soul of higher learning.

I have tried to show in this chapter why a scholar should always supplement the technical task of interpreting the unambiguous meanings of a map with the mastery of a model's multiple, often less determinate, meanings. A competent instructor may have to explain the proper relation of the cartographer to his maps and of the cabinetmaker to 'the different kinds of wood' he makes into furniture or the intuitive feel he displays for 'the shapes slumbering within' driftwood, say, or pine (Heidegger, 14). But to contribute to the conversation of mankind, the scholar should also acquire a third and more important habit. He has to respect and try to emulate the philosopher or historian who masters the great transforming models of his discipline. Martin Heidegger believes that, instead of becoming a famous specialist or expert in his field, a professor should imitate Socrates by 'being more teachable' than his apprentices (71). When identifying the source of the Nile or the height of a mountain peak, a map-maker has no tolerance for ambiguity. By contrast, the model of an atom in quantum mechanics or of a molecule in chemistry has less in common with a map of the Rockies than it has with the syntax of a language or the code of the human genome, whose complex uses are multiple and indeterminate. Whereas a map contains instructions for its own interpretation, a physicist who shares the cartographer's passion for precision has to decide for himself how precise he can expect to be. It may take years to chart the properties of light or the mysterious behaviour of a subatomic particle. The qualities of such a particle are elusive: often its very location is uncertain. Similarly, 'in a dialogue of Plato – the *Phaedrus*, for example, the conversation on Beauty – can be interpreted,' according to Heidegger, 'in totally different spheres and aspects.' Instead of discrediting the thinker's passion for accuracy, a model's 'multiplicity of meanings,' says Heidegger, 'is the element in which all thought must move in order to be strict thought' (71). When the location of an atom becomes a statistical probability or the historical Jesus a multiple layering of different portraits, it is less important to be precise about a given detail than to be precise about the necessary limits of precision in general.

Heidegger maintains that the greatest paradox of modern culture is that even scientists who have mastered complex models of physics and biochemistry have not yet undertaken the more complex task of thinking. When he accuses science of failing to think, Heidedgger is not implying that physicists and chemists are incapable of representing ideas symbolically. He does not deny that they can conceive causes and effects and even forge strong chains of logical demonstration. By failure to think, Heidegger means that scientists have not yet attempted for their disciplines what Socrates attempted for ethics and theories of knowledge when he asked the question: what is the essence or defining attribute of virtue, holiness, or wisdom?

The 'most thought-provoking' of Heidegger's claims is that even 'in our thought-provoking time' science itself 'does not think and cannot think' (6, 8). For it lacks the power to discern the attributes of its own technological processes and what their defining essence might be. It takes a modern Socrates like Heidegger himself to perceive that in its passion for precision science has made an idol of the kind of 'one-sided and one-track thinking' that reduces everything 'to a univocity of concepts and specifications' (32, 34). 'We should fall victim to a disastrous self-deception,' Heidegger warns, 'if we were to take the view that a haughty contempt is all that is needed to let us escape from the imperceptible power of the uniformly one-sided view' (34). To counter the weird, unearthly things that a technological culture spawns, universities must sponsor thinking that is more difficult and radical than mere problem solving. Socratic truth is genuinely ambiguous. Its only proof is an appropriate response, which is less an isolated act of logical demonstration than an apprenticeship or a vocation, a lifelong habit or 'call to think.'

9

Socratic Mentors: Proving Truth by Living It

The secret of Socratic education is that there are two paths to knowledge – through *elenchus* and through embodiment. Truth is disclosed not just through Socrates' exposure of logical fallacies in an adversary's arguments. It is also achieved through an embodiment of wisdom and self-knowledge in the life of Socrates. In the next three chapters I have decided to treat Socrates' second path to knowledge separately, because I find it difficult to travel the path of Socratic mentors at the same time I am tracing the more conservative path of scholars who may minimize the role of bold conjecture in their work.

The three concluding chapters concentrate on the importance of 'a living, spontaneous energy within us,' a form of 'implicit reason,' as Newman explains in a sermon preached on St Peter's Day, 1840, which allows the mind to feel its way swiftly and delicately toward the truth. 'The mind ranges to and fro and spreads out and advances forward with a quickness which has become a proverb, and a subtlety and versatility which baffle investigation' (1887, 257). As the prior and more intuitive activity, the scholar's implicit reason bears the same relation to the real assent of Newman's religious believer as a logician's explicit reason bears to the notional assent of a theologian. Newman accepts Edward Hawkins's principle that to learn religious doctrines we must consult the formularies of the Church. The Scripture's function is not to *teach* these doctrines, which are implicit in the sacred texts, but to *prove* the doctrines by showing how they are embodied in the probation or proving ground of life itself. Doctrine bears the same relation to the lived or enacted truths of Scripture as notional assent bears to real assent.

Many great teachers and prophets are the visible conductors of a Muse or a daemon. Since a wire never generates the electric current it carries, we can no more expect a scholar to explain the secret of his energy or current than we should expect Homer to give a theory of Homer or Newton an insight into his scientific genius. Scholars who prove the truth by living it, who embody the doctrines they cannot explicitly formulate, manage to absorb into their intellectual habits and practice the original, half-instinctive activity by which the mind takes possession of a difficult or complex truth. Only later may these same scholars use logic or formal demonstration to make the process of appropriation self-conscious and articulate. Emerson says that 'proverbs, like the sacred books of each nation, are the sanctuary of the Intuitions' (1908c, 81). We might say the same of aphorisms, which are half oracle and half proverb, a spontaneous overflow of powers which are the best part of a writer precisely because he has no direct or conscious knowledge of them. In one of those precious forms of aphorism which are as close to being perverse as profound, Yeats says we can embody our deepest truths but never know them: 'For wisdom is the property of the dead, / A something incompatible with life' ('Blood and the Moon,' 4.7–8). A scholar cannot at the same time live the truth and understand it.

To assess Socrates' legacy, this chapter briefly reviews the nature of Socratic education in a liberal democracy and the use of the celebrated Socratic method of inquiry. Socrates' criticism of authority and respect for tradition survive powerfully as an academic ideal. But however worthy of theoretical respect, the Socratic method as a practical procedure of mutual inquiry and response has been dead since the invention of printing in the fifteenth century. If we are looking for Socrates' ghost among today's sophists, we may find it at work among a few enlightened scholars and deans. But his spirit survives chiefly in those rare mentors who love what they contemplate and embody what they teach.

In her chapter on 'Socratic Self-Examination' in *Cultivating Humanity,* Martha Nussbaum identifies several important features of a Socratic education. Pluralistic and liberal, it is designed for all human beings and suited to vastly different capacities and needs. She argues that Socratic inquiry ought to be the nucleus of liberal education for four reasons. In the first place, an inquiry capable of withstanding critical scrutiny assumes that a rigorous examination of one's beliefs is necessary for every citizen of a democratic state, and not just for a privileged few. If all citizens are to participate in politics, then universities should follow Socrates' example of recognizing everybody's humanity by trying to educate and talk to everyone. In the second place, just as Socrates imitates the skilful speaker of a dramatic monologue who adjusts his questions to his auditors, so Socratic education in a pluralistic society should be carefully adapted to each 'pupil's circumstances and context' (32). To meet the students' needs and explore the diverse interests that engage them, a school should experiment with 'many different curricular approaches' (32). A third requirement is that Socratic education should accept Mill's challenge to conduct multiple 'experiments of living' that honour the dignity of difference by including a wide spectrum of 'norms and traditions.' Or in Nussbaum's pithy phrase, 'Socratic inquiry mandates pluralism' (33). In the fourth place, Socratic education honours thinkers, not because they are 'museum pieces of intellectual history,' but because their thought comes 'to us as something in each case unique, never to be repeated, inexhaustible' (Heidegger, 75–6).

Since a democratic political culture requires citizens to be critical of the state and detect errors of judgment when they occur, an education that is genuinely Socratic should never invest any professor with infallible authority. On the contrary, schools in a democracy must resist every attempt to turn Aristotle, Newton, or even Socrates himself into the fourth person of the Trinity. Like Socrates, teachers must challenge King Nomos by encouraging students to think through the arguments in great books, making the mind 'more subtle,' 'rigorous,' and 'active' (Nussbaum, 35). Any veneration of authority will 'lull pupils into forgetfulness of the activity of mind that is higher learning's real goal' (34).

Nussbaum has collected data to show that philosophy courses committed to the Socratic ideal of rigorous critical thinking have been successfully introduced in a wide range of American colleges, including Randolph-Macon College in Virginia, Notre Dame, Pittsburgh, and Harvard. She argues that by being trained to 'reason critically in a Socratic way,' students in a liberal democracy are being well equipped to preserve 'the health of democratic freedoms' and 'defend' what she calls 'the democratic value of Socratic citizenship' (49).

Whereas Nussbaum's Socrates is the custodian of freedom and a pillar of liberal democracy, Allan Bloom and Northrop Frye perceive a vastly different Socrates. Their hero is a rebel, a radical critic of state authority and power, even when he makes it a point of honour not to seem so. Bloom and Frye believe that (as a prototype of the modern professor) Socrates placates King Nomos by perpetuating a conservative myth about learning while secretly challenging society with revolutionary alternatives. Far from being the soul of the modern university or even the repository of Plato's own secret tenets, which he shares with only a few select disciples, Socrates' teachings in *The Republic* are a mere protective camouflage. They are a bait to appease Cerberus, the watchdog of custom and authority. The university's true function is not to support or pacify the state but to foster criticism and dissent. By nourishing intellectual or artistic passion, the professors' undeclared mission is to rescue future scientists and scholars from the business schools and from the all-consuming commercial ambitions of the suburban middle class.

In *The Closing of the American Mind* Allan Bloom endorses Leo Strauss's teaching that Socrates is a secret agent or fifth column – a kind of Trojan horse – dragged illicitly into the city he wants to capture. Bloom's Socrates can charm the state partly because it 'is blind to what is most important to him' (Bloom, 1987, 283). By speaking ironically, at a playful distance from his audience, he can protect the 'basic fragility' of his mission. Because the universities speak with a forked tongue that sends a conservative message to the establishment and a more radical coded one to their disciples, their profoundest legacy is masked. As Bloom says, even in pacifying the politicians Socrates leads his most astute and discerning disciples outside the city 'to the Elysian fields where the philosophers meet to talk' (1987, 283). Just as Robert Lowell believes that *Paradise Lost* and *The Divine Comedy* are partly hermetic poems in which Milton and Dante mean at times the opposite of what they say, so Bloom suggests that Socrates challenges the values of the civic authorities while taking care not to appear subversive. Like Milton the rebel, who secretly sympathizes with the revolutionary Satan, or like Dante, who is a soul mate of the radical Manfred, a Ghibelline liberator, Socrates appeals simultaneously to two different audiences.

In the subversive spirit of Allan Bloom, Northrop Frye is less concerned with the Socratic method of answering a question with another question than with the philosopher's habit of being elusive and of holding explosive meanings in reserve. It is less important that the teacher engage the student in conversation than that he function as a kind of drug pusher who can blow the student's mind. As a Socratic ironist, the professor should try to disappear so that something important in the subject can come through. In defending irony, Frye himself becomes an ironist, a modern Socrates who hovers 'furtively on the outskirts of social organization, dodging possessive parents, evading drill-sergeant educators and snoopy politicians' (1988, 20–1). Like the Plato of Leo Strauss, Frye discerns behind the state's official apologist a thinker more revolutionary than anyone suspects. Society should not be allowed to fathom the radical depth of Socrates' mission, which is the immortal task granted only to teachers, the task of corrupting its youth.

But if conservative thinkers like Strauss and Bloom have mistaken Socrates' ironic mask for a face, radicals like Frye may have misinterpreted the face itself. Since each could argue with some display of logic that there is only *one* Socrates, there would seem at least to be *two*. Even if the double nature or duplicity of Socrates' teachings is only a fiction, it is a fiction Socrates has carefully constructed and helped popularize. Indeed the astonishing facility with which thinkers from opposing schools have managed to kidnap Socrates (and claim him as a kindred soul) strongly hints that, far from being the advocate of a single system or idea, Socrates is a truly protean thinker. He is a kind of Shakespeare among the philosophers, with no single method or doctrine to expound.

To engage with this protean Socrates is to see how the scepticism and caution that pull the mind several ways at once accurately reflect the uncertainties and doubts of a culture caught between two worlds, 'one dead,' as Matthew Arnold says, 'the other powerless to be born' ('Stanzas from the Grande Chartreuse,' ll. 85–6). As I argued in *Origins of the Monologue: The Hidden God*, the antithesis or tension of opposites imparts a keen sense that the search for a centre of spiritual authority eludes the grasp of many speakers in post-Enlightenment monologues. The old faith has grown intolerable; it is undermined and doomed; but the hunger for something new is a genuine *cri de coeur*. Like theologies asserting that God is simultaneously absolute, infinite, and a first cause, the soul of higher learning remains a mystery. In his changing use of the word 'idea' in his great book on education, even J.H. Newman (I think) tries to put a headlock on an eel. Like Kant's God or the properties of light in modern physics, the 'idea of a university' is as about as stable and luminous as an *ignis fatuus* or phantom. To compass any of higher learning's great issues is to cast a net of contradictions. Unless learning is interdisciplinary, it lacks a context, and so collapses into a heap of isolated specialties. But until each specialty also coheres around a disciplinary model, like iron filings round a magnet, no inquiry can take place. In other words, a genuine exchange of ideas is at once necessary and impossible.

A similar logic dictates that a profession that is not liberal is not professional, and that a liberal education in philosophy or history that fails to build like Euclid from axioms to theorems is not liberal. A specialist who knows only his own subject knows not even that. For to know its territory well he has to know the countries that lie just across its borders. And to explore these countries he has to be the opposite of a specialist, a learned amateur. Such antinomies or 'feuds,' as Marjorie Garber calls them, between academic disciplines, amateurs and professionals, or jargon and plain language, are always unstable. But since Garber believes 'many of the contrasts around which these disputes revolve do not signify real opposites,' she argues that they 'depend upon one another for their strength and effectiveness' (x). It seems to me, however, that any defence of jargon or any apology for the ascendancy of a single scientific theory like evolution, which nineteenth-century thinkers impose on everything from theology to ethics, is ridden with sophistries. I am as sceptical of Garber's unmanageable optimism about the 'truth of opposites' as of Gerald Graff's unhelpful advice to 'teach the conflicts.'

Just as successful teachers can embody or live their most important truths without consciously professing or even knowing them, so education, I think, can promote culture most effectively by declining to mention it. To be too explicit about one's goals is to make two mistakes. It is to confuse the conscious means of liberal learning with its more important, less self-conscious purpose or goal. And it is to assume erroneously that the truths most worth imparting can be planned in advance and consciously transmitted. When the means is mistaken for the end, the aims of education may have to be boldly redefined in the manner of A.N. Whitehead's opening salvo in *The Aims of Education*: 'culture is activity of thought and receptiveness to beauty and humane feeling,' he polemically asserts in an overflow of high spirits. 'Scraps of information have nothing to do with it. A merely well-informed man is the most useless bore on God's earth' (1949, 13). The professor who is too self-conscious about his goals may confuse the clearing of hurdles with intellectual growth. It is possible to acquire knowledge easily but be stupid in the only sense that finally matters. For as T.S. Eliot warns, 'culture can never be wholly conscious – there is always more to it than we are conscious of; and it cannot be planned because it is also the unconscious background of all our planning' (1948, 94).

When a professor tries to embody or live the truths he cannot formulate in a set of rules, his dialogue may have more in common with the tips shared by a coach than with the traditional mentoring of a tutor or an academic lecture. According to Donald Schön, coaching has three essential features: 'it takes place in the context of the students' attempts' to practise their profession; 'it makes use of actions as well as words; and it depends on reciprocal reflection-in-action' (101). One reason why professional education is expensive is that the mutual inquiry and response of coach and student cannot take place in a large lecture hall.

It must occur in the intimacy of a seminar room, a laboratory, or a hushed conversation between a senior physician and a medical student by a patient in a hospital.

Whereas the discourse of demagogues is below verbal expression, the aphorisms of Socrates and Jesus are above it. Their ability to speak without saying anything converges on moments when flint meets steel. In his words from the Cross, as in his Sermon on the Mount, Jesus so thoroughly inhabits a universe of earlier biblical texts that what he says is one part speech to three parts silence. After citing the opening line of Psalm 22: 'My God, my God, why hast thou forsaken me?,' Jesus can, in one commentator's words, 'triumphantly [skip] the rest of that Psalm – and [say] nothing. But the silence speaks.' Because Psalm 23, with its exalted poetry of peace and consolation, 'is vibrating there in the air around him,' Jesus can make his listeners 'fill in the silence, if only mentally, with what comes next' (Hart, 103). The words of great teachers reach into silence. In encountering the shock of life-and-death issues, they often replace logical coercion with wonder or persuasion.[1]

Emily Dickinson says that 'the unknown is the largest need of the intellect' (quoted by Kazin, 155). Since the little we know is a mere shadow of the magnitude we shall never know, higher learning is often an attempt to establish with Socrates or Kant the limits of the knowable. According to Vico's formula 'verum factum,' the only truths humanists know for certain are truths they themselves have fashioned. But there are several reasons why scholars remain unconscious of these truths. Sometimes the truths they speak die in the heart. G.B. Shaw quips that 'the unconscious self is the real genius. Your breathing goes wrong the moment your conscious self meddles with it' (281). True scholars and trailblazers from Socrates to Nietzsche may also lack a vocabulary to name what is most original in their thinking and destined to survive them. Since a portion of the *terra incognita* that scholars are mapping is also the act of mapping itself, they often know more than they *know* they know. Their repressed knowledge of the truths they embody must sometimes be recovered for them by intellectual historians or by critics operating as the psychoanalysts of culture.

Scholars and scientists can educate students to think clearly and recognize errors when they occur. But they cannot teach invention or any deep and subtle method of discovery. Just as Aristotle claims that the creation of metaphor is a feat of synthetic imagination that cannot be taught, so the talent and genius of prophets, rebels, and poets blow like the wind, wherever they will. Though we can identify an area of knowledge like mathematics where utility and abstract pleasure are likely to coincide, it is impossible to identify in advance the precise sciences – physics, genetics, molecular biology – in which a generation's greatest discoveries will be made. Moreover, the mystery of knowledge is nowhere more evident than in its contradictory requirements of learned dilettantism and the expertise of generalists. In an age of increasing specialization, in which chemistry subdivides into biochemistry and physics into quantum mechanics and nuclear physics, each new discipline needs highly technical investigators. It is equally true to say, however, that the progress of human knowledge also depends on the harvest within a single mind of ideas gleaned from many disciplines.

T.S. Eliot claims that 'skepticism is a high and difficult faith.' There is even a sense in which scepticism is not so much unbelief as a diffident way of expressing belief. Though Newman never doubts the mysteries of his faith, he is diffident or sceptical about the power of his words to express them. He is an 'intelligent man,' as Owen Chadwick says, 'who knows that the conscience does more for truth than the intelligence; an eloquent man who fears oratory, a subtle logician with the lowest opinion of logic; a fertile mind with small regard for originality' (225). For Newman, as for Hopkins, religion is an incomprehensible certainty rather than an interesting uncertainty. Any theology or intellectual system withholds from such minds as much as it reveals.

The abiding themes of great literature and art are almost never what they seem to be, for their most important ideas are what Hopkins calls their 'under-thought.' Philosophers who embody their truths, as Socrates does, cannot directly know them. And if scholars and scientists talk too unreservedly about their methods, they will give away secrets and substitute a theory of their discipline for a dedication to its practice. Yeats speaks of the artist's 'secret discipline,' because he knows that artists who best exhibit the painter's 'stern colour' and 'delicate line' ('In Memory of Major Robert Gregory,' 9.3–4), which are a product of long apprenticeship and practice, are seldom bold enough to talk about these qualities openly.

'Tell all the Truth,' says Emily Dickinson, 'but tell it slant' (lyric 1129, l. 1). Since a truth expressed is usually a lie, Cordelia in *King Lear* stubbornly refuses to commit the civil falsehoods of her sisters. Instead of lying to tell the truth, she realizes that what a daughter feels for her father she does not express. As Edward Tayler admits, the power of *King Lear* cannot be explained: 'Nobody can lay a glove on this play. This is the greatest thing written by anyone, anytime, anywhere, and I don't know what to do with it. In a case like this, no one else knows what to do with it, either' (Denby, 305). Literature is just as cognitive as science, but it transmits a different form of knowledge. It replaces the sequential arguments of ordinary reasoning with the presentational forms of a parable, a witticism, a joke, or a metaphor, whose force must be grasped all at once or not at all. *King Lear* transmits not just what Keats calls 'the holiness of the heart's affections and the truth of the imagination' (257). The secret of its art is also its truths of embodiment, which it offers as an alternative to truths of logical demonstration and rebuttal. What can be shown but not said in great literature is unteachable, because it consists of questions we must live with, not of problems we can solve like a puzzle.

Whenever I try to talk in a lecture or seminar about *King Lear*, I lose the object. Unlike Christ's suffering, a tortured redemptive suffering like Lear's or Cordelia's *cannot* overcome death. In contrast with a traditional tragedy of martyrdom, which strengthens the community of faith, the play's arcane message is unsettling and subversive. But something I cannot understand or explain brings the tragic drama to the point where its conventions nearly disintegrate yet mysteriously do not. Iris Murdoch ingeniously suggests that 'tragedy must mock itself internally through being essentially, in its own way, a broken whole' (116–17). Like Cordelia, however, who refuses to violate the primal silence of the family, which feels but cannot express its love, I think there are times when a critic should quietly stay with Shakespeare's words and forgo explanations. In Wittgenstein's aphorism, 'there are, indeed, things that cannot be put into words. They make themselves manifest. They are what is mystical' (1961, p. 151). In an aspiring scholar the love that is most enduring is often disappointed. For to fall in love with such imaginary objects as a definitive lecture on *King Lear* or an authoritative monograph on mystery is to love an object so elusive that it must fail to satisfy the whole longing of the scholar. Like the pursuit of rainbows, it is a love without possession.

In Yeats's great poem 'The Folly of Being Comforted,' the speaker is unable to decide whether he is inconsolable or not in any need of consolation. As the woman he loves grows older, her 'great nobleness' and grace of self-command make 'the fire that stirs her when she stirs / Burn but more clearly.' But how can the 'nobleness' of age compete with 'all the wild summer' of her youth, which suddenly flares up in the poet's memory like flame from a collapsing funeral pyre? Though capable of being separately appreciated and appraised, the two kinds of fire are incompatible: nothing allows the poet to choose between them. The mystery comes, not from something unknowable in the subject, but from something unlimited in it.

For similar reasons Robert Frost insists on keeping secret the values he most wants to share with his readers or impart to his students. 'For God's sake,' he tells his friend Sidney Cox, 'never let the word Responses loose in a classroom. Have them in all their variety but name not their dread name' (1981, 201). Only in the privacy of a letter to Cox can Frost's secret thoughts about secrets be intimated. Ironically, what Frost remembers is that his most important thoughts elude him. He cannot quite formulate or understand what he wants to say. Regrettably, Frost's thoughts on education surface only randomly in his letters. They remain a mere torso, like Cox's own projected book *Winking at the Sphinx*, the insights of a literature professor that Frost keeps urging his friend to commit to paper and share.

For a memorable seminar discussion in which the unsayable seems at times to manifest itself we can always turn to the famous feast of discourse in Plato's *Symposium*. But there are very few transcripts of stirring seminar exchanges or lectures. When Northrop Frye and Vladimir Nabokov publish their lectures as books, the wit and charm of their public performances tend to be as insubstantial as the pageant that fades in mid-air in *The Tempest*. Lucy Winter, the main character in May Sarton's novel *The Small Room*, thinks that since inspired teaching comes only 'from immense inner reserve,' its mystery has 'to be as carefully guarded as the creative power of the artist' (117). The drama of shared ideas can seldom be staged beyond the charmed circle of the lecture or the round table of the seminar. The magic is fugitive and dies with each class.

Many years ago in California I met a brilliant teacher and scholar who told me in confidence that his performance in the lecture room was merely off-stage acting. It was like one of the masques prepared for Miranda's entertainment by Prospero. When he was not at the university teaching Blake and writing books about him, he spent his time (not learning Sanskrit or poring over obscure treatises on mysticism) but watching stock-car races at the Riverside speedway or TV sports events in local bars. To this day I cannot decide whether he was an imposter or a genius who knew that to live one's scholarly life too intensely is to play the teacher's role badly.

Scholars may presumably take off their masks in the presence of graduate students and peers. But if teaching undergraduates is make-believe and theatre, then every professor should aspire to be an actor. For a scholar this is surely the sublime and refined point of felicity, as Swift might say, the pleasure of performing for novices, the dazed and mindless state of being a deceiver of innocents.

Like my California colleague, who confessed to leading a double life, many gifted teachers take their secret art to the grave with them. One great teacher who shared secrets, however, was Reuben Brower, who confided to his Harvard teaching assistants that the best ideas in his seminars and lectures were spontaneous and unrehearsed. Though few professors can conduct a lecture the way Robert Frost conducts a poem, which unfolds by surprise, the moments one remembers are often lucky combinations of careful planning and impromptu asides. A satisfying lecture is like an elegant solution to a mathematical puzzle: its arguments are simple but subtle, involving some combination of fulfilled expectation and surprise. There was never a more cerebral lecturer than Northrop Frye. Yet a quip like 'the Romantics invented the copyright laws' seems to have sprung unbidden to his mind as he made his way to the lecture room. Despite its neoclassical propriety, even T.S. Eliot's aphorism – 'The lesser poets borrow, the great ones steal' – retains the spark of an unrehearsed aside. The spark of invention, of course, may detonate its materials as well as ignite them. The greater the interval the spark must jump, the greater the risk of producing sublime folly rather than sense. But there is also a greater likelihood of intense illumination, as in a metaphysical conceit that is both surprising and just.

As if to intimidate me or throw down a gauntlet, the organizers of the Priestley Lecture at the University of Lethbridge told me that the year before I gave my own talk Barbara Hardy had delivered an immensely successful lecture without a single note. The committee was even more astonished when, before she flew back to Britain, Professor Hardy presented the committee chairman with a manuscript that repeated every word of what had sounded the night before like a spontaneous address. Surely more remarkable, however, than such prodigious feats of memory, which merely *simulate* thought in process, were Christopher Ricks's Alexander Lectures at the University of Toronto, for which no transcript existed. A few hours before each lecture the organizers sequestered Professor Ricks in a seminar room with copies of the poems he wanted to discuss. At the beginning of each lecture Ricks composed himself to silence then spontaneously launched into the most searching and exact analysis of poetic allusion I ever heard.

Though it is better to commit a lecture to memory than to read it from a script, it is best for someone who has thought profoundly about a subject to follow the Muse's lead and think aloud. Despite his great gifts as a writer, Herman Melville was a poor lecturer who bored his audiences by gluing his eyes to a script. By contrast, Emerson's lectures (like Frost's) appeared to be spontaneous. In bringing his 'puppet show of Eleusinian mysteries' to the populace, Emerson spoke like an oracle whose inspired phrasing seemed just as startling to the seer himself as to his audience. One of the great masters of impromptu lecturing was Sigmund Freud. To simulate the intimacy of a two-sided conversation, Freud would pretend to address a single member of his audience. He would even interrupt his lecture to engage in a Socratic dialogue with this person, and would be thrown into confusion if she failed to appear. When asked by Ernest Jones a few minutes before a lecture what his subject would be, Freud replied, 'If I only knew! I must leave it to my unconscious' (cited by Kevin Jackson, 288).

According to Mortimer Kadish, friendship is one of the great neglected models of higher learning. To be a student of Socrates we must first become his friend, since he lives his truths like Jesus and embodies them in his parables and sayings. Though what is most worth learning cannot be taught, it can be imparted through embodiment and example. Perhaps Socrates' method justifies the education in silence and suffering that Rabbi Saunders inflicts on his son Danny in Chaim Potok's novel *The Chosen*. The mentor as father or friend teaches habits of mind and ways of living that disciples must unconsciously assimilate and love before they can consciously imitate or try to understand. But just as God has no friends, so a professor who wants to share ideas has to descend from his podium by substituting inquiry for pronouncement and dialogue for decree. I have found that most students yearn for ties of personal influence. As if 'the wounds left by his own Oxford experience were still open,' Newman cries out many years later: 'I have known places where a stiff manner, a pompous voice, coldness and condescension, were the teacher's attributes ... This was the reign of Law without Influence, System without Personality' (quoted by Cameron, 56–7). It is true that for a few adventurous souls the university can be a place of lifetime exploration, where mentors and their students map new countries of the mind. But such journeys of discovery are rare exceptions, not the rule. A professor of philosophy told me in anguish that none of his three daughters, all in their mid-twenties, had ever known the influence of a great teacher.

In *Out of Place: A Memoir*, his depressing recollection of his graduate instructors at Harvard, Edward Said recalls Newman's account of 'an academical system without the personal influence of teachers upon pupils.' Its 'arctic' winter creates 'an ice-bound, petrified, cast-iron University and nothing else' (Newman, 1856, 256). Said clearly thought of I.A. Richards as one of Harvard's great men, someone who could concentrate in his lectures and person the virtues of a whole university. But what Said remembers about his fallen hero is enough to discourage any scholar: 'Twice I tried to study under the aging ... Richards, the most avant-garde figure then at Harvard, and twice he defected from his courses just after midpoint, when his secretary would enter and say that the course had been unilaterally dissolved. He was a comic miniaturization of the once-adventurous thinker – vague, vain, rambling – and as I read it, his major work struck me as thin and unaccomplished, as unprovocative as Blackmur's was stimulating and, despite its gnarled syntax, suggestive' (290).

An anecdote or quip is often the best way of remembering a teacher's special qualities. After falling off a dais, George Lyman Kittredge observed that for the first time he found himself on an equal footing with his students. And upon receiving an armchair presented to staff members after twenty-five years of service, Howard Mumford Jones summed up his restiveness by noting that 'Harvard is, so far as I know, the only institution in the country that says to employees of long standing, "I think you had better sit down"' (Bate et al., 66). Though his intellect was just as formidable, Douglas Bush was a gentler, more diffident teacher than Jones. Committed in equal measure to a sense of fair play and to a distaste for dissension or ill will, Bush told C.P. Snow that his book *The Masters* was 'the only thing [he] ever read that *completely* reconciled [him] to Harvard's system of autocratic appointments' (Bate et al., 87). Like the tail that wags the dog, the memorable tale still dogs the waggish mentor or don.

Unlike the flamboyant Santayana, who climbed out a window to escape the embarrassing attention given Harvard professors in their last lectures, the more reclusive Bush escaped applause by unobtrusively finishing his courses a week before the end of term. Whereas Bush avoids altogether the disturbing ceremony of a closing lecture, Simone Weil's mentor, Alain, meets his class two days after the last scheduled session, which was too cluttered with dignitaries, he felt, to give charity and justice the attention they deserved. 'No adieu,' as George Steiner says. 'Grandeur has its reticence' (107).

A senior scholar once told me that the crown of his career was a festschrift he received to honour his retirement. But such an honour must surely signal that the hour is at hand when he will be active no more. I tend to share Collingwood's suspicion that 'the last humiliation of an aged scholar' occurs when 'junior colleagues conspire to print a volume of essays and offer it to him as a sign that they now consider him senile' (119). Instead of being presented with festschrifts or giving valedictory speeches at retirement parties, departing professors should follow the examples of Bush and Santayana. After talking to close friends over sherry and sharing a few memories in private, they should pass quietly out the door. Just as an older generation of mentors once greeted the retiring professor, so he should welcome the young instructors pressing past him through the door of a profession that is closing behind him but which he would probably enter again gladly if he had another life to live.

Closer to Socrates than we are, the New Testament celebrates the intimate bond between the mentor and his disciples in such acts as the touching of a garment's hem and the ritualized washing of feet. Though Anglo-Saxon societies are too repressed to permit the kind of physical touching encouraged in some pseudo-Platonic dialogues, everyone understands the force of touch as a metaphor: I was touched by his kindness. Even after 'we have been rightly alerted to all forms of sexual harassment,' as Seamus Heaney says, 'there can be such a thing as vocational harassment, where the student's hopes and aspirations are unthinkingly assailed' by a cold and distant teacher (72). To be in touch with their students, great teachers cannot be too Olympian, or 'too encased' in what Heaney calls 'the shining armor of *moi*' (72). Though Dr Arnold's sermons at Rugby were 'clear and stirring as the call of the light infantry bugle,' Tom Brown never feels that his mentor's 'warm' and 'living voice' was speaking to him 'from serene heights.' On the contrary, his mentor was his friend, an ally constantly 'fighting' at his side (Hughes, 143–4).

Though inspired teaching still remains a mystery, despite all the wise and foolish books that have been written about it, Northrop Frye believes it is usually a five-stage odyssey, ranging from bashful self-absorption at the beginning to an impersonal openness to vision at the end. In the first stage the instructor 'is an embarrassed medium, of limited personality, reading carefully from notes and trying to let the subject reveal itself in the clarity and patience of the exposition' (2000, 17). At the second stage the mastered notes are a mere stage prop to give the lecturer a sense of security as he looks up timidly to address a few diffident students. At the third stage the university lecturer, grown more confident, may turn into a prism rather than a window through whom the light of knowledge passes without refraction. In Frye's words, he becomes a 'Covering Cherub, an opaque black priest, or rather preacher, putting on a personal show for the benefit of [his] pride' (2000, 17). At the highest stage teaching once again acquires impersonal authority. As the whole room fills up with the informing spirit of Shakespeare, Milton, or whoever is being taught, the students behold something larger than themselves or something behind the lecturer that 'accounts for his being there' (2000, 17).

One mystery of good teaching is the power of some professors 'not only to teach us to think,' as J.H. Newman says of his mentor Richard Whately, but also 'to think for [ourselves]' (1968, 22). Though no teacher can be taught how to do this, good teaching, like good scholarship, often raises more questions than it answers. Instead of forcing the learner's pace, it is also a product of reflective leisure. To an ample subject like Spinoza's *Ethics* it will allot an ample run of time. For this reason the best lectures and books often seem to be written out of knowledge that goes far beyond their given subject matter. Any pressure to teach more or larger classes or to cover a topic more rapidly is in danger of turning the university into an intellectual boot camp. It loses touch with the old Greek meaning of the word 'scholastic,' a devotion to learning of one's leisure, of one's 'time to be wise.'

The distance separating the profoundest mentor from the shallowest instructor must seem to the former trifling and to the latter infinite. Though I have had only one influential mentor in my life, F.E.L. Priestley, I do not understand to this day the source of his influence. Priestley's ironic, withering lucidity often intimidated me. But strangely enough, he became an intellectual father to me. I was shocked – and initially resentful, I think – to discover at his memorial service that my mentor had a nearly identical effect on many other students. Perhaps the puzzle of an influence that seemed as unique as friendship or love can be partly explained in retrospect. As a Newton scholar and a philosopher of science as well as an authority on Tennyson and Browning, Priestley combined a scientist's understanding of intellectual history with a poet's gift for irony and two-way meanings. Even as he explained difficult ideas in concepts common to everyone, he seemed to be speaking (as a poet speaks) to each of us alone and waiting for our personal reply.

Whatever else inspired teaching may be, it is not the soliloquy of a genius communing alone with God and the prophets. If we recognize our own thoughts in a memorable seminar, it is not because they are ideas we once rejected that now 'come back to us,' in Emerson's curious phrase, 'with a certain alienated majesty' ('Self-Reliance,' 1908c, 33). Great teaching is more like a Socratic dialogue in which the mentor, always an attentive listener, is a midwife assisting at the birth of our own ideas. This is why Oscar Wilde says that, though 'education is an admirable thing, ... it is well to remember from time to time that nothing that it is worth knowing can be taught' (1968, 206).

No commentator on education is more insistent than J.H. Newman on the need for close association with a mentor. But few commentators claim to be more perplexed why truths must be embodied in a lecture or tutorial rather than merely expounded in a book. In *The Office and Work of Universities*, Newman is convinced that apprentice scholars must consult the master 'and listen to his living voice.' But he is hard pressed to say why this is so. 'Whatever be the cause,' he confesses in some bewilderment, 'the fact is undeniable. The general principles of any study you may learn by books at home; but the detail, the colour, the tone, the air, the life which makes it live in us, you must catch all these from those in whom it lives already' (1856, 13).

I think that Newman's own Oxford sermon on 'Implicit and Explicit Reason' (1840) provides the clue he is seeking. Lectures and tutorial instruction are important because 'implicit reason' is inseparable from the living, spontaneous energy of the scholar who possesses it. A student must perceive this energy at work in the modulations of a lecturer's voice, in the dramatic vividness and authority of a mentor's look and bearing. What 'baffles investigation' in Newman's Oxford sermon is the mind's power to seize truth by a 'half-instinctive' process. In *The Office and Work of Universities* that bewilderment transfers itself to the power of great scholars and teachers to embody in their own unique endowments, by practice rather than by rule, the power of instinctive reasoning.

Newman's distinction between two kinds of reason, explicit and implicit, corresponds to Socrates' distinction between two paths to truth: rational inquiry and embodied wisdom. It also anticipates Newman's well-known distinction between two kinds of theological assent, 'notional' and 'real,' in his *Essay in Aid of a Grammar of Assent*. When a master of the art of 'implicit reasoning' wants to learn something new, he should ask if he can teach it. For to scale a new peak of knowledge is to exercise an inward faculty that is best displayed in the performance of a gifted teacher. 'To men in general,' however, as Newman warns in his Oxford sermon, the exercise of this faculty constitutes an expedition of the mind that is 'as unsafe and precarious ... as the ascent of a skilful mountaineer up a literal crag' (1887, 257).

Without profaning a mystery, some qualities of implicit reasoning can be dimly discerned in the practice of a great mentor. When I took Priestley's Victorian and eighteenth-century courses at University College, I felt like Keats's astronomer when a new planet swims into his ken. Since Priestley, like Socrates, held meanings in reserve, as if further disclosures had to await new and better questions, I could sometimes detect a flicker of disdain or irony in his asides. Partly because Priestley never referred to notes, I soon began to realize that his lectures were occasions where thought was in process. Though my mentor has been dead twenty years, I still reach out to him. I often think of Priestley as Frank Kermode thinks of his friend and mentor Peter Ure: 'I continue to admire him and even, in some privileged moments, feel his presence half admiring, half deploring whatever I am trying to do at the time. I still have difficulty in imagining anybody so different from myself who might nevertheless find something in me to like' (173). Authentic teaching is a calling. It summons disciples, not just to the terraces of a Tower of Babel, where professional students scramble furiously for monetary reward, but to pinnacles of Parnassus. Unlike the gravediggers, whose pedantry is a tomb, the best teachers are touched by the divine spark. They use the grace of their obsessions to open a path to the Muse.

With the lofty aplomb of Moses on Mount Sinai, it is sometimes decreed that all great teaching aspires to the condition of Socratic discourse – a process of mutual inquiry and response. I am suspicious of this dogma on the ground that such a process of inquiry is difficult to sustain, except among intellectual equals, and is too time-consuming to practise. Socratic conversation is usually a monologue in disguise. Typical of the many advocates of Socratic discourse who disagree with me is Mortimer Kadish, who replaces Michael Oakeshott's favoured metaphor of the conversation with a ballet in which the 'dancers, faculty and students alike, become competitors who need' each other's input (Kadish, 64). Preferring to pass at will from the formal lecture in a large hall to the intimacies of a conversation in a college common room or to the sharp exchanges in a seminar, Kadish's scholar-teacher is a flexible pragmatist who refuses to prescribe a single model of learning. He recognizes that a genre may mask its true identity: a lecturer's monologue may be dialogical, an interplay among multiple voices, like the intellectual wrestling that animates the Psalms or John Donne's sermons. And even a Socratic dialogue like the *Phaedo*, when sufficiently authoritative,

can be 'monological' in its single-minded efforts to impose on its participants a theory of recollection or a doctrine of immortality. Willing to
concede that a gifted mentor or teacher should combine all three
genres of instruction, Kadish believes that 'the moral of the Socratic dialectic for higher education' lies in just such an interplay. 'Protagoras
questions Socrates, Socrates Protagoras; [and] each finally must question himself' (64).

The most sustained attack on the Socratic dialogue as the privileged
genre of educational discourse comes from O.B. Hardison. Implying
that the Socrates of the magisterial dialogues is an argumentative bully,
Hardison is afraid that 'trying to maintain the Socratic posture leads
teachers to coercion and intellectual hypocrisy.' The cogency of his
assault depends on his first assimilating to the magisterial Socrates the
self-critical Socrates of the dialogues of search. Hardison's Socrates is
covertly authoritative, a knowledgeable guru in disguise. But since
'nobody knows all the answers,' the modern Socrates who aspires to
omniscience always 'make[s] a fool of himself' (137).

In Hardison's polemic the so-called Socratic dialogue 'degenerates
into a series of mini-lectures.' In a less common scenario, the instructor
may take risks by imitating Socrates' exploratory method in such dialogues of search as *Lysis*, *Charmides*, and *Laches*. Since the Socrates of
these dialogues no longer knows the answers to the questions he asks
about friendship, temperance, or courage, the discussion moves 'in circles.' 'As the students realize what is happening' in such exchanges,
'they become bored,' Hardison maintains, and begin to 'withdraw mentally' (137).

According to Hardison, however, aporetic discourse is theoretically
improbable, an open-ended inquiry that seldom occurs in practice.
Most Socratic discourse is either openly or secretly doctrinal. It manipu-
lates its participants by prodding them 'to some preordained conclusion.' Hardison fears that the result of such prodding 'is a system based
on falsehood, a con game. Like antigravity, it is antipedagogy. In place
cf free human communication, you get a master-slave relation that
tends to corrupt' both the lazy tyrants who masquerade as teachers and
their 'apathetic or rebellious' subjects (137).

In 'Crisis in the Classics,' M.I. Finley replaces Hardison's coercive
model of indoctrination with a persuasive Socratic model of two-way
conversation. The Greeks remind us that 'the whole idea of inculcation
is at fault ... All reading, all education,' he insists, 'is a dialogue; what
one puts in is as important as what the other side contributes' (Finley,

15). On balance, however, Hardison's objections to the concealed coercion and hypocrisy of most Socratic discourse seem to me unanswerable. Outside the aporetic dialogues of Socrates and some dramatic monologues by Browning, Socratic discourse is rarely found in practice because it is too risky to implement and too time-consuming to sustain.

When I read the testimonials written on behalf of a successful teacher by former students, I receive a strong impression that these students (now barristers, medical doctors, or professors themselves) are eager to repay an important debt. They feel grateful for being in contact, at an impressionable period of their lives, with a mind gifted for experience of intellectual adventure. Like Socrates and other teachers whose contributions are not in the main reducible to writing, such professors may not be remembered when the generation of students they have taught passes away. But on rare occasions a disciple who has been graced by a mentor's educated taste or refreshed by his innovative teaching may record his debt. Encounter with genuine learning may seem as improbable as winning the lottery. But like travel to a strange land, it is always a surprise.[2] Regrettably, no Socrates can count on having a Plato in his class to keep alive his memory. But just as the church needs its prophets as well as its priests, so a mentor who wants to be remembered by posterity may need to associate himself with a few protégés more gifted than he is. As Harold Bloom sadly warns, 'time, which destroys us, reduces what is not genius to rubbish ... If there is secular immortality it belongs to genius' (2002, 814). Who would remember John the Baptist without Jesus or Leopold Mozart without his son, the immortal Amadeus?

The qualities required of a good administrator are just as elusive as those of a prophet, psychiatrist, or music coach and almost as rare.[3] In general, scholars and teachers have most confidence in deans and chairmen who possess some talent for scholarship themselves. A successful administrator is more than a bookkeeper or a politician. He must also be a diplomat who openly discusses policies he wants to promote. Like kings of a country with a fragile economy, always on the brink of collapse, deans should never bribe a citizen to counterfeit the currency. One administrator has repeatedly asked me to rewrite my letters of recommendation by using rote gestures and bogus formulas. But any scholar attuned to language can tell at a glance when a recommendation is bloated, anemic, or flagrantly dishonest.

There is a comic half-truth to Clark Kerr's quip that 'the three major administrative problems' of a university president are 'sex for the students, athletics for the alumni, and parking for the faculty' (quoted by

Charlton, 19). Without the professors' right to contemplate, question, challenge, and discover, which deans must support, universities will soon become old age homes for the faculty, except that more people will die in colleges than in retirement homes. A wise chairman will also encourage conversation among a department's aloof but famous scholars, who tend to dwell on different peaks of Olympus.

It is appalling that eminent scholars at the same university, who live together in the same community, should seldom if ever meet to discuss their ideas. In his *Autobiography* R.G. Collingwood pays handsome tribute to F.H. Bradley, whom he calls the greatest of the English idealist philosophers. He even defends Bradley from the unfair criticisms of their Oxford colleague Cook Wilson, who attacks Bradley for views the older philosoher never held. But then comes Collingwood's astonishing disclosure: 'although I lived within a few hundred yards of [Bradley] for sixteen years, I never to my knowledge set eyes on him' (16, 19, 22). Such excess of English reserve seems quite uncivilized, refined beyond the point of decorous behaviour. Deans and masters of colleges should do everything in their power to bring luminaries like Bradley and Collingwood together.

A good chairman will also respect and try to reward the occasional Socrates who refuses to expound his deepest thoughts in writing because he fears they will be valueless out of context. But chairmen should also possess enough wit to humour reticent scholars with something to say into publishing books, even when they lack the authority to intimidate or shame them. Deans and masters might do worse than circulate the joke Collingwood tells about Cook Wilson. 'Whether [Wilson] thought that by not publishing he deceived the public into thinking that he never changed his mind, and whether he regarded this as a good thing to do, even though the public remained ignorant what his mind was, or whether he had a mind at all, I did not ask' (19).

Just as the Sabbath was made for man and not man for the Sabbath, so the protocols of deans and provosts are made for scientists and scholars, not scientists and scholars for the deans. The decision of Victoria College to make Northrop Frye its principal is like a church's choosing Thomas Aquinas as its pope. To reward an intellectual achiever by elevating him in the hierarchy is to honour merit by squandering a rare resource. If the university is a casino of enlightened self-interest, each gambler (including provosts and deans) should take a course in the investment and use of rare commodities.

Though university presidents and deans tend to think of education as a business enterprise, scholars and research scientists are likely to be most productive when accepting the working fiction that they pursue knowledge because they want to. The idea that the university is an entrepreneurial business corporation compromises those scholars who are in love with their discipline and want to enter their profession the way a bridegroom or a bride enters a marriage. F.H. Bradley says that marriage is 'a contract to pass out of the sphere of contract' (1876, 174). If a scholar is married to his discipline, his positional interest is surely to be non-positional, to be as disinterested in the pursuit of truth as Matthew Arnold's 'saving remnant' in *Culture and Anarchy*. By turning the faculty into a proletariat, the collective bargaining favoured by some deans and chairmen often betrays another mystery: the special bonding between professors and their students. To create walls and boundaries between faculty and students or between faculty and management is, in James Cameron's strong but eloquent words, 'to abandon the entire university tradition and to accept the term *university* as denoting an employing authority, like Bell Canada or Imperial Oil' (78–9). 'For the faculty to negotiate with the "university" is to admit that the faculty is not the university, to abandon the notion of the *universitas*, to accept a proletarian status, not under protest, but as a norm' (78).

Most merit schemes for professors have all the disadvantages of a superstitious faith in business methods without any of the advantages of religion. The increases are large enough to spread the heresy that diverse academic talents can be quantified. But they are too small to inspire belief that the chairman who administers the rewards is a just and loving Father. Like faithful members of a business corporation, professors at the University of Toronto who receive the largest increases are required to 'surpass expectations.' Heard in one tone, this appeal is a mere empty echo of every scholar's private resolve to be Ulysses: 'to strive, to seek, to find and not to yield.' But heard in another tone, this call to surpass is also a seductive siren song. It is a temptation to exchange the ever-present risks of a challenging project in scholarship or science for the ease of lotos land. It encourages an adventurer in ideas to seek the security of a lucrative contract with a publisher or a predictable and safe short-term payoff. Though it is better to be a clone of Jeremy Bentham, a James Mill or G.H. Lewes, than a scholar *manqué* like Casaubon, in the retrospect of history it is also better to be an experimental and self-critical thinker, an unpredictable Utilitarian like John Stuart Mill, than any of his brethen. A prolific robot more concerned to surpass the expectations of chairmen and deans year after year than to meet his own expectations is unlikely to satisfy the expectations of posterity.

Deans and chairmen who administer a mindless system of merit increases that reward the quantity of research rather than the quality of scholarship do more to depress than energize their faculty. By ranking professors in the popularity contests that too often pass for accurate assessments of their teaching, the meritocracy encourages the very competitiveness, aggression, and breaches of decorum that defeat civilized conversation and prevent learning from being 'an ornament in prosperity,' as Aristotle says, 'and a refuge in adversity.'

There can be no conflict between the authority of wisdom and knowledge, because in transcending knowledge wisdom 'never contradicts or humiliates it,' as Frye says of reason and revelation (1964, 306). A university's spiritual authority depends, not on its deans, presidents, and famous research institutes, and not even on the university chairs, which are too often endowed by the academic equivalent of a political party or a lobby. It depends rather on the presence of a few rare thinkers, often on the margin of politics, where they hope to avoid the warfare of their peers. Arnold calls this saving remnant the 'aliens,' persons 'who are mainly led, not by their class spirit, but by a general *humane* spirit, by the love of ... perfection' (Frye, 1964, 316). Though a centre of spiritual authority exists, it is often (as Arnold realizes) 'the place of greatest isolation' and risk (Frye, 1964, 316).[4]

The mystery of what makes a good scholar is as baffling as what qualities are necessary in a distinguished administrator or teacher. There are too many scholars *manqués*, superfluous veterans who, never having spent their mental capital before retirement, try to be productive as they near the grave, resolved to outlive themselves. At the other end of the spectrum, I am alarmed to discover how often the academic system has destroyed the soul of higher learning by rewarding the pedant who can cite all the authorities and conventional arguments while penalizing the rare Socrates who thinks for himself. Many graduate schools have filtered out intellectual risk-taking and adventure. Knowing enough to be dogmatic but not enough to be original and just in their assessments, young scholars try to be savants rather than seers. If forced to toil too long in the vineyard as poorly paid teaching assistants or tutors, scholars may lose their passion for learning and ideas. There is some truth to Oscar Wilde's observation that 'everybody who is incapable of learning has taken to teaching – that is really what our enthusiasm for education has come to' (1968, 166).

Though the system's survivors are diligent jumpers, skilled at clearing hurdles, they seldom make bold and speculative thinkers. A doctoral student who feeds off the lifeblood of his mentor is half vivesector of the mentor's reputation and half vampire. Scholarship owes more to the errors of a learned but freewheeling rebel than to the correct opinions of a mental slave. A flinty maverick ought to strike sparks of fire from hard resisting data. If a young thinker is intellectually alive, he will see his world with the eye of awakened imagination. With scornful lucidity, he will flash light into the dark corners of a subject. But seldom does an inspired guess set fire to the mind of an apprentice scholar. Seldom does a great gamble make an obscure topic clear. Since there are no rules for winning in the lottery of scholarship and science, life in this uncertain world might seem too unsettling and inconclusive – were it not for the fact that zero risk and guarantee of success are precisely what no intellectual explorer can possess or should seek. To avoid risk is to fall back into a master-servant duality that empties scholarship of energy and will. It turns thinkers into slaves who fear the freedom of a bold idea. It produces robots who lack the spark of invention, who reject the challenge of contributing to knowledge, and who commit the ultimate treason of renouncing the demanding but liberating high adventure of a life devoted to thinking for oneself. Though it is hard for Dante the pilgrim to say farewell to Virgil in the *Purgatorio*, every scholar must outgrow his mentor in order to become himself. A good student becomes a bad scholar when he is unable to survive his master.

10

Prophet, Rebel, Poet: The Scholar's Hidden Knowledge

If Aristotle is right to praise metaphorical speech as a sign of high intelligence, perhaps we can recognize the Socratic wisdom of prophets, rebels, and poets in their talent for aphorism. Like Socrates and Jesus, many prophets and scholars are masters of wise or witty sayings. Often they join the wit of a discovery to the judgment of a verdict or a law. The wit dissolves the authority of an accepted truth, and the judgment gives a revolutionary new oracle the ring of an encyclical decree or a law received from God. Wallace Stevens's rabbi in 'The Auroras of Autumn' realizes that, unlike the phrase 'An unhappy people in an unhappy world,' which provides 'too many mirrors for misery,' the aphorism 'An unhappy people in a happy world' gives the prophet the proper syllables 'to roll' on his tongue. Before the 'secretive syllables' of a great homily can be 'solemnized' ('The Auroras of Autumn,' 10.1, 3, 6–7, 12), it must possess the aesthetic effect of an oracle or a prophecy. A teacher like Northrop Frye is probably least Socratic in those moments when he forsakes logic and exposition for oracular aphorism. In his own words, 'the lion, unlike the monkey and the squirrel, is not a conversationalist, and the rhetoric of authority is leonine rhetoric: "The Lord will roar," as Amos says (1:2)' (1982, 212). But I think a teacher who combines Jesus' talent for aphorism with Socrates' attention to evidence, logic, and experiment can surpass even the oracular Frye by becoming a lion and a monkey, a prophet and a scholar, simultaneously.

The typical probative aphorism, which captures the aphorist in the act of discovery, is witty rather than judgmental. An example is Oscar Wilde's aphorism: 'A Truth in art is that whose contradictory is also true' (1968, 157). If this aphorism is true, it would seem to follow it is also not true, for the same reason that the statement 'I am a liar' is false if it is true and true if it is false. Conversely, once 'the pleasures of sudden wonder are ... exhausted,' a magisterial aphorism can 'repose' with authority 'on the stability of truth,' as Samuel Johnson both affirms and illustrates in *The Preface to Shakespeare* (1958, 241). Another example is Johnson's own definition of wit, which combines stability and repose in equal measure. True wit rotates securely on an axis of truth, whose poles are the new and the natural, the surprising and the just. Equally imposing and self-assured is Kant's categorical imperative: 'Act as if the maxim of your will were the law of the universe,' which translates into the Golden Rule: 'Do unto others as you would have them do unto you.' As soon as George Bernard Shaw adds his witty coda – 'Don't do unto others as you would have them do unto you: their tastes may be different' – a magisterial aphorism becomes probative (2002, 18). Shaw's aphorism invites us to conduct an experiment of living to discover what another teacher or student might choose in our place. Whereas the Decalogue is magisterial, the Sermon on the Mount is probative. Pope's aphorisms are often only witty aperçus. But in the fourth book of the *Dunciad* they take on the air of fearful last words, spoken just before the prophet moves into darkness. Though just as daunting, the aphorisms of Blake and Nietzsche stake a claim to revolutionary truth. Whatever its syntax, the deep grammar of a magisterial aphorism is descriptive; and the grammar of a probative aphorism, which may substitute the paradoxes of the gospel for the maxims of the law, is performative – a way of doing something new with words.

Though heuristic aphorisms are usually witty and magisterial ones judgmental, a gifted teacher may sometimes reorganize these groupings. An aphorism that is probative and exploratory may acquire the force of a judgment if it is stated as a creed or an axiom, like Blake's 'Auguries of Innocence': 'To see a World in a Grain of Sand / And a Heaven in a Wild Flower / Hold infinity in the palm of your hand / And Eternity in an hour' (ll. 1–4). If an experiment is successfully performed, then its discovery of the macrocosm in the microcosm can be celebrated as a universal truth. In the hypothesis of Tennyson's poem 'Flower in the Crannied Wall,' the speaker is asked to conduct an experiment: '*If* I could understand, / What you are, root and all, and all in all, / I should know what God and man is' (ll. 4–6). The italicized conjunction dramatizes the impossibility of completing the task. But if the whole experiment could be conducted, the discovery would be proportionately great. The aphorism has the judgmental force of a prophecy. But despite the fact that it proposes an experiment only God

himself could oversee, the aphorism is less magisterial than exploratory and heuristic.

By the same token, a teacher's magisterial aphorisms can be witty
rather than judgmental. Bernard Shaw is a master of aphorisms whose
authority is no less magisterial for being wittily parasitic on some other
aphorism: 'Do not love your neighbour as yourself. If you are on good
terms with yourself it is an impertinence; if on bad, an injury' (2002,
18). Shaw's aphorism feeds destructively off the second Great Commandment, which retains the normal judgmental form. Sometimes a
single fertile saying will breed a whole incestuous chain of witty aphorisms. When Wilde quips that 'one touch of nature may make the whole
world kin, but two touches of Nature will destroy any work of Art' (1968,
176), he feels impelled to invent a third aphorism to tell us why. A sunset painted by Turner is at first so unique that Nature, forgetting 'imitation can be made the sincerest form of insult,' keeps on repeating the
effect till no critic of taste can endure another sunset (1968, 188).
Unfortunately, when predictably witty aphorists like Shaw and Wilde
cease to shock us, they often cease to be interesting.

An aphorism that is witty often uses words in an altered sense. Examples include Hopkins's tortured line, 'I wretch lay wrestling with (my
God!) my God' ('Carrion Comfort,' l. 14), or Donne's punning self-
reference in 'having done that, Thou hast done, / I have no more' ('A
Hymne to God the Father,' ll. 17–18). A judgmental aphorism, by contrast, may offer a solution to a sonnet's problems that is epigrammatic
and concise: 'all / life death does end and each day dies with sleep' ('No
worst, there is none,' ll. 13–14). But often an aphorism that seems judgmental may start to effervesce the moment we place it in context. At
first glance no aphorism sounds more authoritative and judgmental
than Pope's 'Hope springs eternal in the human breast' ('An Essay on
Man,' 1.95) or Shakespeare's 'One touch of nature makes the whole
world kin' (*Troilus and Cressida*, 3.3.175). But restore these aphorisms to
their contexts and the wit starts to erode their claim to be universally
true. Equally witty and grammatically active are the capping couplets of
many Shakespearean sonnets, as Helen Vendler has shown.

Often a judgmental aphorism becomes witty when a prophet or sage
allows its emotive or descriptive meaning to change. In Carlyle's aphorism that democracy 'means despair of finding any Heroes to govern
you' (1970, 289), both the descriptive and emotive meanings of 'democracy' start to shift subversively. In Blake's aphorism 'Pity would be no
more, / If we did not make somebody Poor' ('The Human Abstract,'

ll. 1–2), the altered emotive force of 'pity' accompanies revolutionary new insights of the kind Heathcliff proposes when he rejects Nelly Dean's appeal to 'pity,' 'duty' and 'humanity' in *Wuthering Heights* (Brontë, 125, 128–9).

Polonius's speeches in *Hamlet* illustrate the mind-numbing effect of foolish aphorisms. As Coleridge observes, Polonius resembles many pedants in his embodiment of a wisdom no longer fully possessed. Sometimes foolish aphorisms are self-refuting, like the long-winded Polonius's aphorism that 'brevity is the soul of wit.' Often an addiction to aphorism is also a way of holding life at two removes. Such is the effect of Mr Duffy's elaborately contrived aphorisms in Joyce's story 'A Painful Case.' There is a similar aesthetic disdain for the contagion of ordinary life in the aphorisms of Walter Pater, Oscar Wilde, and Henry James, especially in *A Sense of the Past*. An aphorist like Frost's farmer in 'Mending Wall' may even use an aphorism about good fences making good neighbours to fence *out* more meaning than he is capable of fencing *in*.

Wise folly is a signature of the great aphorisms of Shakespeare's Fool in *King Lear*. The aphorisms of Blake, Nietzsche, and Yeats's Crazy Jane often affirm the essential sanity of prophetic genius and the madness of more commonplace minds. By repeating a mantra – 'Shantih, shantih, shantih' – an entranced yogi relaxes his tensions and passes out of time into a time-free state. More ambiguous is the folly of T.S. Eliot's wise man in 'Journey of the Magi.' Like the aphorist as seer, he senses a numinous meaning lurking either *in* the natural world or *beyond* it. If *in* the natural world, the seer's genius is sacramental, and he tends to write beautiful aphorisms. If the hiding place of power is the invisible world, the seer is mystical or visionary, and he coins aphorisms that are sublime rather than beautiful. Traherne's verbal jewel – 'the corn was orient and immortal wheat, which was never reaped nor was ever sown' (*Centuries of Meditations*, 3:3) – is an example of the beautiful or sacramental aphorism. Vaughan's vision, 'I saw eternity the other night / Like a great ring of pure and endless light' ('The World,' ll. 1–2), is more mystical.

Sublime aphorists are always on the threshold of insights that no words can quite compass or express. There is often an inspired form of nonsense in their utterance: 'Oh, God! It is unutterable! I *cannot* live without my life! I *cannot* live without my soul!' (Brontë, 139). Though tautology is in principle the tritest trope, it is potentially the most powerful. Because the words inscribed on the Temple of Isis conceal more than they reveal, Kant celebrates them as the most sublime ever written:

'I am all that is and that was and that shall be, and no mortal has lifted my veil.' Shelley's aphorism that 'the deep truth is imageless' (*Prometheus Unbound*, 2.4.116) is equally sublime. But the epitaph Pope composes for Newton's tomb in Westminster Abbey – 'Nature and nature's laws lay hid in night; / God said, "Let Newton be!" and all was light' – is magisterial and witty. Though its couplet reposes on an axis of truth, its pretence of knowing God as well as God knows himself is in sharp contrast with Newton's own picture of himself as a little child collecting only a few pebbles on the seashore, while the whole ocean of knowledge lies unexplored before him. Pope sometimes substitutes a 'wise saying' for an insight. The delivery of less meaning than is promised produces a merely trite aphorism. Even a profound aphorism may become trite when its author is deficient in invention or wit.

Frye maintains that 'the cautious legal cough of parenthesis has no place in a prophetic style.' By contrast, paradox and sublimity are the hallmark of oracular aphorists whose 'every sentence is surrounded by silence' (1982, 212). There is a touch of oracular sublimity in Carlyle's 'Natural Supernaturalism,' which is composed almost entirely of aphorisms: 'Man is a minnow, his Creek this Planet Earth' (236), or in the discontinuous prose of Thoreau's *Walden*: 'The light which puts out our eyes is darkness to us. Only that day dawns to which we are awake. There is more day to dawn. The sun is but a morning star' (824). Such aphorisms help the intellectual explorer glimpse through the arch of experience the gleam of Ulysses' untravelled worlds whose margins fade for ever and for ever as he moves.

Most visionary or mystical aphorisms are sublime; and most sacramental aphorisms are beautiful. But occasionally the ratios reverse themselves. Combining lyric excellence with the lapidary concision of an aphorist, Tennyson's dramatizations of the spirit and wind in sections 86 and 95 of *In Memoriam* are steadily and luminously visionary. But they are also beautiful rather than sublime because of their exquisite evocation of local English landscapes. Conversely, despite the 'wildfire' that leaves only 'ash' in its wake, Hopkins's Incarnational language in 'That Nature is a Heraclitean Fire' is potently sacramental: 'In a flash, at a trumpet crash, / I am all at once what Christ is, since he was what I am' (ll. 20–2). The tidal surge and flow of the seascapes in *The Wreck of the Deutschland* are tempestuous and ravaging rather than pastoral and serene. Yet even the heroic, Job-like language of the nun's baptism of 'her wild-worst / Best' (24.8) evokes a sublime sacramental world always on the verge of fragmentation and collapse. Such unstable oracles often induce vertigo.

Other examples include Walter Benjamin's saying that 'every finished work is the death mask of its intuition,' or Santayana's aphorism: 'There are books in which the footnotes, or the comments scrawled by some reader's hand in the margin, are more interesting than the text. The world is one of those books' (quoted by Kevin Jackson, 284, ix).

The aphorism or oracle is the preferred genre of the scholarly browser, who after reading a text from beginning to end may want to trace a 'path that has no ends and no beginnings, where all is middle.' Such is Geoffrey O'Brien's strange account of browsing, which he describes as a 'sort of hermetic spiraling, no more static than the vibrant stillness of the hummingbird at the window.' To browse is 'to amble through a sentence as if it existed independently of any other sentence,' like a detached aphorism or an oracle, and 'to allow its implicit rhythm to pervade all surrounding space' (64).

Since the aphorism or parable is the preferred medium of many great prophets, rebels, and poets, including Socrates and Jesus, I devote attention to it in this chapter. As a presentational use of language, an aphorism must be apprehended as a whole, not analysed into parts like sequential discourse. Iris Murdoch devises a test to distinguish presentational from discursive forms. She argues that even a bad pun or aphorism is still a pun or an aphorism. But a defective geometrical proof, consisting of a series of illogical operations, disqualifies itself as a proof. Unlike sequential thinking, presentational language can be imperfect without ceasing to be itself.

The best aphorisms are located somewhere between the two extremes of platitude and paradox. If an aphorism veers too far in the direction of a puzzle or a paradox, it turns into an unintelligible riddle. And if it goes too far in the other direction it turns into a truism. Some aphorisms may be precipitated out of a crisis of understanding, as when a truism dies so that a new law, like the kerygma in the Gospels, can arise as a phoenix out its own cremated body: 'You must lose your soul to save it,' 'The last shall be first,' and so on. Without antithesis ('The last shall be first'), there can be no spark of wit, and hence no room for paradox. And without chiasmus ('The last shall be first and the first shall be last'), there is little opportunity for elegant reversal.

A paradox can be epigrammatic as well as oracular. But an epigram that is merely chiasmic is unlikely to seem paradoxical or profound: 'Those who matter don't mind, and those who mind don't matter.' Though the hint of an antithesis may lurk in the flickering contrast between 'mind' and 'matter,' it exists only by virtue of a half-pun or anti-pun which the epigram evokes as a mere ghost or shadow. What distinguishes not merely the aphorism but wisdom itself from platitude is the gnomic wit of the expression.

Aphorists are always wrestling with the necessary angel of continuity. Since they are artists in search of a medium that is continually eluding them, their aphorisms are a pledge or promise of more wisdom to come. An aphorism is a bookmark: a slip of paper inserted into a text or discourse we want to return to, an insight we hope to revive. Though it is seldom a spot of time, a moment made luminous, a marker can identify a site worth revisiting – a place where 'the light that never was, on sea or land' once broke forth and may appear again. The more proverbial or epigrammatic an aphorism, the more it tends to enforce some tenet by redeeming a platitude: the more paradoxical an aphorism, the more it tends to generate new insight and perception.

In casting mere shadows on the dial, the gnomist as gnomom will conceal more truth than he reveals. His aphorisms may be subtly off-key, as John Bayley says of the most aphoristic lines in *King Lear* ('Ripeness is all,' for example), which try to compass a meaning that will not quite formulate. The aphorisms of Browning's Andrea del Sarto and even of Keats at the end of 'Ode on a Grecian Urn' ('Beauty is truth, truth beauty') are comparably off-key, because they evade subtle truths that the speakers discover in the course of the poems. But even in their odd impertinence the aphorisms may be psychologically revealing. In their very irrelevance they may be luminous. Aphorisms may enforce an old code, or they may invent a new one by conducting an open-ended experiment in living, as Jesus does in the Gospels. A foolish aphorism is the grave of genius; a wise one is its temple; and a sublime aphorism is the chrysalis at the moment of release, when a winged spirit is about to take flight.

Since even prophets correct their proofs, I have tried to show how the secret knowledge of seers, rebels, and poets can sometimes be analysed in carefully constructed aphorisms. Like the many samples of elliptical, half-formed sentences used by psychiatrists and music coaches that Donald Schön has studied, aphorisms supply important documentary evidence. A prophet who is witty but not oracular may degenerate into a buffoon. A prophet who is oracular but not witty may be only a mouthpiece of platitude. A merely profound prophet may be too obscure to have either authority or wit. One part speech to three parts silence, a powerful aphorism is the outward sign of a genuine prophet, the trace of a departed genius or daemon.[1]

Though retreat behind the walls of a discipline is often the first refuge of ignorance, a measure of academic purity may be necessary to prevent the hidden knowledge of an aphoristic seer – a prophet, rebel, or poet – from being kidnapped by a momentarily ascendant discipline like computer science or statistics. Just as universities in the Middle Ages had to combat the heresy that all learning aspires toward the condition of theology, so modern universities have to fight the heresy that all knowledge aspires toward the condition of a practical science. There are many cells and cloisters in 'the house of intellect,' and if we knock down too many walls the roof may collapse.

As a humanist who has always admired and often learned from my brother, a professor of physics, I have tried hard to imitate what is boldest and most enduring in the sciences. Though I reject all reductive explanations that put God or ethics in a test tube, I honour in Newton and Darwin what I honour on a smaller scale in Northrop Frye: a power to see occult resemblances among things apparently dissimilar. What William Whewell, the great Victorian philosopher of inductive logic, calls a 'consilience of inductions,' an intuitive leap that allows dissimilar classes of facts to be 'jumped together,' reaches across all the sciences to inform many of the branches of humane learning treated in this study. Recently rediscovered by Stephen Jay Gould, 'consilience' is a feat of logic that allows Northrop Frye to discover genuine analogies between Gothic fictions like *Great Expectations* or *Jane Eyre* and the Cinderella story. In Newton, Whewell admires the elegance and beauty as well as the astonishing explanatory power of a single doctrine of attraction, which 'encompasses,' as Gould says, 'all three of Kepler's previously uncoordinated laws' (Gould, 210). As a humanist I value in Darwin the same simplifying power of consilience that I value in theorists of irony like Kenneth Burke and W.H. Empson, who draw into a single web of sense such seemingly disparate threads as Sophoclean irony and the irony of fate, or Socratic irony and the irony of a double or two-way meaning.

In the tradition of Walter Redfern's monograph on puns and Eleanor Cook's lively study of word-play and word-war in Wallace Stevens, the best scholarship and criticism often tap hidden or unconscious forms of knowledge. The best insights often withhold more meaning than they impart, and like a joke or a pun they have to be understood all at once or not at all. By shocking us into new awareness (as in the Beatitudes or Shaw's *Revolutionist's Handbook*), a witty scholar like Northrop Frye encourages us to ponder the mysteries and live the questions instead of merely answering them as part of an intellectual exercise.

Because the most important truths are often hidden and have to be ambushed or taken by surprise, Socratic education is more likely to be an improvised expedition than a planned journey. Every scholar knows the excitement of wandering down an unexplored bypath or apparent detour that turns out to be a main highway. Even an obscure footnote may lead a scholar to his next essay or book. In good conversation, as in good scholarship, Hans-Georg Gadamer explains that 'something comes out that neither of the partners contains in himself' (Perkins, 1988, 116). Such an experience may take place when a scholar finds his own indecision mirrored in the self-divisions of the philosophers or poets he is studying. We are by instinct monists and by instinct dualists. 'The soul is naturally Christian,' says Tertullian; and it is also naturally pagan. Indeed each mind is so deeply divided that 'four thousand volumes of metaphysics,' says Voltaire, 'will not teach us what the soul is' (both quoted by Dixon, 24).

As I argued in the last chapter, a Socrates who embodies authentic qualities of mind and spirit will not stoop to lecture about these qualities, nor will he openly allude to them. Instead, he will give his auditors a chance to overhear these values in his conversation or observe them in his life. And only by developing the habit of discerning a personal intelligence in natural events and the discourse of peers will scholars and scientists acquire intelligence themselves.

There is an element of irreducible mystery or secrecy in complex mental operations, because in such operations judgment (or 'knowing how') is increasingly required to supplement the simpler requirements of information (or 'knowing that'). Judgment plays an important role, not just in historical scholarship or literary criticism, but also in what Oakeshott calls 'the unspecifiable art of scientific enquiry, without which the articulate contents of scientific knowledge remains unintelligible' (1989, 55). The further we progress beyond simple technical and motor skills, the less easily our knowledge can be processed into more information. We have to *interpret* the facts and rules. And where a conflict between rules exists, it can seldom 'be resolved by the application of other rules. "Casuistry," as it has been said, "is the grave of moral judgement"' (1989, 55).

The seat of higher learning is not just the university we graduate from but the university we carry in ourselves. And its soul 'is both mysterious and substantial,' as Frye says, 'infinitely beyond us and yet inside us, something we can never reach and yet something that is essentially what we are' (1988, 92). A genuine student moves in a circle of committed thinking. And true commitment is itself an involuntary or unconscious act. We must feel about it what Frost says he feels about a poem he may write. One is shy to speak about it in advance, since it comprises a series of lucky events and unfolds by surprise as it goes. Conscious choices are less a result of deep commitment to a vocation than of casual guesses about the market value of commerce versus physics, for example, or about the skills we perform with most dexterity or ease.

Too much conscious planning and reliance on archival sources may destroy a book. Preferable to the stress of working in a large library or archive, where no excuse exists for not being erudite, is the unconscious knowledge recollected in a mood of tranquillity or wise passiveness that allows a scholar to reflect on what he has read and see more clearly with the 'inward eye' of memory. Written during the war with only a small select library at his command, Eric Auerbach's *Mimesis* is a masterpiece. But when the same scholar taught at Yale, 'the vast resources of the Sterling library' ironically proved a mixed blessing, as Alvin Kernan says: 'the result of riches was a dreary book on rhetoric, read and used by few' (1999, 108).

Some harsh truths may have to be forgotten or suppressed if professors are to teach year after year and still survive. The ultimate failure of even the most inspired teaching is that, unlike some scholarship, it is ephemeral. As Julius Getman explains, 'the most profound feeling of failure that I recall came, paradoxically, after a particularly successful first-year contracts class ... There was no record, no lasting result – nothing but an hour of excitement that I realized would soon fade from everyone's memory' (42). Even more common is a professor's anxiety or fear that the next lecture or seminar will invite student indifference or disdain. The personal interaction may be missing: the students may dislike the professor's persona. They may want to be entertained by a spellbinding performer or awed by an acknowledged expert rather than challenged by an open-minded risk-taker or led by an explorer. Last week I was elated by two reviews of a book I had written. But two days later an *ad hominem* attack on my teaching in the Anti-Calendar brought me to a new low. Getman observes that often professors have to pass within the cycle of a single week from 'academic success and universal expression of esteem to failure and student disdain without understanding what is happening to them' (42).

I was recently elated when a philosopher I admire told me he enjoyed our conversations on Hardy and Schopenhauer in the college lounge. A mere two months later, however, I was depressed beyond measure when he told me at his retirement party that he had never counted me among his colleagues or friends, because I had failed to make myself a presence in the college. Though I value conversation, I find ideas are generated more spontaneously around the college photocopy machine than by the coffee urn in the senior common room or among the tea cups in the lounge. And yet his remark was biting and hurtful; it went quite deep, I suspect, because it had the sting of truth.

I still have nightmares of the outdoor class I gave over forty years ago as a substitute instructor at Wellesley College, when the notes I had carefully planted away in my copy of Arnold's poems were blown down a hill into a lake. Equally distressing was Walter Houghton's directive that, since some Wellesley students had seen me arrive by public transit, I must either walk or take a taxi to future classes. Unless the professor is prepared to live with daily recognition of his failures, he may have to forget about satisfying students or colleagues and try to satisfy himself instead. To this end he should be very wary of reinventing his persona by giving up daring or experimental roles for safer conventional ones.

For four decades I have been trying to discover why scholarship and teaching are at once fulfilling enterprises and exercises in frustration. I suspect the fulfilment has something to do with the glimpse academic life sometimes offers of large, enhancing goals. Without an ideal of health, justice, or redemption, the medical doctor, the lawyer, or the theologian would have little motive for inquiry. But I suspect my equally strong disappointment has something to do with the necessary elusiveness of each transforming vision. Frye isolates part of the problem when he says that to enter a profession is at least to make the gesture of 'recognizing the ideal existence of a world beyond [one's] own interests' (1963, 65). Unfortunately, as Frye then goes on to say, 'the civilization we live in at present is a gigantic technological structure, a skyscraper almost high enough to reach the moon.' This modern Tower of Babel 'looks very impressive, except that it has no genuine human dignity.' For all its wonderful machinery, we know that like Enron or the twin towers 'it's really a crazy ramshackle building, and at any time may crash around our ears' (1963, 67).

Another peril of intellectual growth is that it may seriously impede professional advancement: Charles Sanders Peirce was too brilliant and eccentric a philosopher for the permanent appointment at Harvard or Johns Hopkins that his fellow pragmatist William James tried to secure for him. Conversely, the professional solidarity and group support that often help a scholar write a good book may diminish the chance of his writing a great one. In the first case, the risk-taking of the scholar who is prepared to undertake years of unrewarding research before producing a book that is well informed and genuinely innovative may fail to receive the short-term recognition that secures its author tenure. In the second case, the intellectual chance-taker is a victim of the pressure to conform and the dubious rewards of being a Marxist, a New Historicist, or an adherent of some other school. Bewitched by the glamour of an Ivy League college and an elitist ethos, the young scholar finds it easier to produce a stylish or trendy book on cultural history or feminism than an enduring masterpiece like *Anatomy of Criticism*, *Mimesis*, or *The Mirror and the Lamp*.

Julius Getman says he is grateful that his first teaching position after Harvard Law School was at an egalitarian institution, Indiana University. The absence of elitist stereotypes allowed him to develop an aptitude for scholarship that he might not have acquired at Ivy League Yale, where he later taught. At Yale the pressure to conform and publish might have been too inhibiting. One wonders if Marshall McLuhan could have written *Understanding Media* or *The Gutenberg Galaxy* at Cambridge, where he did his graduate work, or whether Northrop Frye could have written his book on Blake at his alma mater, Magdalene College, Oxford.

The fact that professional advancement and intellectual self-growth are often at odds alienates scholars and increases their sense of failure. To secure promotion is sometimes to play it safe, to do what is academically expedient and intellectually conformist. The reluctance to take risks may be professionally useful but it thwarts self-growth. Getman concludes that 'some kind of failure is an almost certain aspect of academic life, even (or, perhaps, especially) for its most creative scholars and teachers' (x). The innovative professor who challenges the assumptions of the cultural historian or feminist critic may be denounced as an enemy. And in refusing to provide students with notes that can be reproduced mechanically on exams, the challenging lecturer who makes students think and refuses to accept undigested data from the website may be denounced in teaching evaluations as disorganized or obscure.

Academic success is often difficult to measure and achieve. A medical doctor is said to succeed if he builds up a reasonably large practice and performs routine operations without killing his patients. He is not required to discover penicillin or a cure for cancer. But gifted scholars and scientists are expected to increase the store of knowledge. As Getman says, their 'goals are far more difficult to achieve than success in many other fields or success as a student, which is the experience most likely to lead people into academic life' (x). Indeed a certain amount of failure goes with the territory. The best book one has written may arouse the envy of a reviewer; even the extra energy one puts into teaching may vex students who resent anything that inhibits their inertia or prods them into active thought. At every level of academic life, success is elusive; and 'failure of one sort or another is always lurking' (Getman, xi).

Getman is troubled by two further anomalies: the more gifted the students, the less they need their professor's guidance. And in choosing a teaching over a professional career, every university teacher in a professional faculty is preparing his students to live a life he has rejected for himself. 'Embedded in this simple fact is an awful irony,' Getman says. 'We seek meaning by preparing students for a life we do not find meaningful' (14). Unless professors are preparing their students for the contemplative life of a scholar or for a scientist's dedication to research, they are like novice masters in a monastery who, after training their students in theology and the rules of their order, find that the novices are no longer fit for life in a monastic community and so must be turned loose again upon the world.

Because most students are conservative and elitist, even when they think they are radical, it is often necessary for professors to reinvent their personae. The worst class I ever taught was a class for fourth-year honours students in my specialty, Victorian poetry. I had just returned from ten years of study and teaching at American universities, and was still experimenting with an informal Socratic method that encouraged students to be quizzical and nonconformist. It is harmless enough for a professor to deceive his students by pretending to be more sceptical or less serious than he is. What is less forgivable is the bad faith that makes professors lie to themselves. Though I affected an unassuming manner and simulated ignorance, Toronto students wanted to be taught by a great man like Northrop Frye or Marshall McLuhan. Despite the radical masks they wore, many of the students were elitist. They loathed democratic professors and preferred to revel in the Oxbridge or Ivy League status of a patronizing sciolist who pretended to know the answer to every question that they asked.

Though the scholar who writes a great book adds to the stature of his discipline, he can seldom transmit to disciples the wit or insight that are his most precious legacy. Northrop Frye writes books of amazing virtuosity, and Stephen Greenblatt is a critic of immense ingenuity and learning. But sooner or later their followers must be made to feel redundant. For to know what myth criticism or New Historicism *is*, they have to

watch what Frye and Greenblatt *do*. Since great scholarship is a practice
not a theory, the one gift that the master who transforms a discipline or
invents a school cannot bequeath to posterity is his most important gift:
his informed sympathy, intelligence, and tact. The wit-criticism that sees
a link between the structure of the Bible and Plautine comedy, or
between Montaigne's report of a marriage between two women and the
plot of *Twelfth Night*, is dextrous and agile. But no school has a monopoly on genius or wit, which seems to emerge unpredictably as an
acquired characteristic of a few scholars and critics with no lasting effect
on the species' 'germ plasm.' As even the rituals of scholarly succession
fall into decay, much of what is most worth transmitting in a learned tradition is never passed on.

Experience proves the utility of both caring and not caring about
short-term results. Nothing can be more practical or useful than nuclear
energy. But nuclear devices depend on the refined speculations of quantum mechanics, whose theories are as arcane to the layman as an
abstraction in metaphysics. As James Cameron observes, 'a passionate
curiosity unaffected by any desire for practical results does in fact produce, without aiming at, results that are useful by the standards of the
world' (15). Unfortunately, a university department of chemistry or
physics that fails to pursue knowledge for its own sake causes the knowledge in that discipline to atrophy or decay. And by emulating the practical success of chemists employed by pharmaceutical companies or of
physicists who design aircraft, the university scientist exchanges the
long-term benefits of pure research for an immediate payoff. In pursuing disinterested scholarship or pure research the university is simply
being itself. Like most other self-justifying activities, its commitment to
knowledge and self-growth 'is unavoidably conservative' and cannot
without compromising its goals 'be jiggled about' like a dinghy 'to catch
every transient breath of wind' (Oakeshott, 1989, 103).

Another great frustration is that the books and projects that most
directly nourish a scholar's growth and most enhance his learning are
often too difficult or specialized to be used successfully at even the most
advanced levels of teaching. I have found to my frustration and dismay
that, despite pious platitudes to the contrary, the demands of instruction and research seldom coincide. On the one occasion I tried to
engage a graduate seminar in the research I was conducting into the
intellectual background of Victorian poetics, members of the group told
me that the work was too time-consuming and difficult. The sad truth is
that some of the best scholars also seem to 'lack the ability to excite,

interest or even educate students about the subject matter' that engages them (Getman, 28). And yet the universities would be incomparably poorer places without them. Even though students find it very difficult to learn from failed teachers, many of these teachers are good scholars who make important contributions to their disciplines.

I am often surprised to discover how even distinguished lecturers and famous scholars have deep misgivings about their teaching. In his diverting memoirs, *Not Entitled*, Frank Kermode recounts a dream that might have been funny were it not so devastating. Kermode imagines he is giving a lecture at which the unsparing Graham Hough denounces all Kermode's writings, walks out of the hall, and is followed by most of the audience. The dreamer ascribes Hough's hostility to the fact that Kermode stayed away from his rival's funeral. But the displeasure of the rest of his audience he attributes to a deeper cause. The dreamer had recently been offering a graduate seminar at Yale, and had a persistent feeling that he 'was performing inadequately and letting [the students] down, though they were far too polite to give any sign that they thought so too.' Then comes the astonishing confession: 'Behind that memory was another,' Kermode admits, 'of my inveterate conviction that I was far from being a good teacher, not an easy thought to have constantly in your mind when what you have principally done to earn a living has been to teach, or profess to' (167). Despite the inventive absurdity and charm of Kermode's dream, it taps a deep level of his psyche and so is not less disturbing for being a dream.

One of Julius Getman's fellow law professors told him, 'I was constantly afraid that someone would come into the classroom and arrest me for impersonating a law professor' (28). The fact that 'many professors never get over this feeling' leads to a tortured lifelong sense that their devotion to ideas is of no direct use or interest to their students. And yet such professors invariably cherish what they themselves apparently fail to provide: the memory of a few favourite mentors who once inspired them. Sometimes the power of such memories to renew commitments and replenish the mind becomes the only way a professor can continue to think and write, week after week, month after month, as a scholar.

As an analytic philosopher with the soul of a humanist, Ludwig Wittgenstein speaks wisely about the hidden quality of knowledge: 'The aspects of things that are most important for us are hidden because of their simplicity and familiarity' (1972, proposition 129, 50^e). 'A great intimacy,' one literary critic says, 'never directly expressed but very

secure,' may signify through a gesture as simple as a 'nod or glance ...
the happy solemnity, the shared acceptance of a poem, even of a line of
verse' (Kermode, 195). Indeed Morton White believes that education is
most successful when indirect instruction in a subject through demonstration or example is substituted for direct analysis of it. A scholar probably learns more about the religious mind by reading Dostoeveski,
Kierkegaard, Browning, or Pascal than by taking a course advertised in a
college calendar as Protestantism 101 or Catholicism 202. We could do
worse than imitate Sir Richard Burton, who keeps telling strangers he is
going to Meccah via Jeddah, although his route is actually Al-Madinah
via Yambu. In justification of this deceit Burton quotes the Arab proverb: 'Conceal thy Tenets, thy Treasure, and thy Travelling' (1:140). Just
as Burton both honours and imitates the intricate evasions of the Arabs,
so the scholar knows that what is most worth discovering is unteachable.[2] Our deepest commitments must be 'picked up' or captured *en
passant*, as J.S. Mill says, while we pursue something else. Like marriage,
discovery of a happy intellectual life is too important a matter to be left
to conscious choice alone.

In May of 2001 the Congress of the Social Sciences and Humanities
Council of Canada concluded that, since Canadians are not a notably
creative people, a commission on creativity should be established to provide a few incentives. One hopes that the commission will be less solemn
than its founding council. For there are surely limits to what any culture
can plan or hope to legislate. Indeed the more education arrogates to
itself the responsibility of reforming or amending culture, the more it
betrays it. Hence the value of scepticism, which T.S. Eliot defines as the
strength to defer a decision, just as pyrrhonism is the incapacity to make
one (1948, 29). Though a culture can daily live and practise its deepest
truths, it inadvertently betrays them whenever it tries to identify or
express them too explicitly.

I want to digress for a moment by considering how T.S. Eliot explores
hidden knowledge in 'Journey of the Magi.' Few students of education
are more aware than Eliot that only a portion of any culture is 'transmissible by education' (1948, 108). And few poets have better expressed the
sense that, even when an important revelation is at hand, it will be difficult to comprehend or put into words. Like many historians, Eliot's wise
man also finds that the more facts and information he collects, the less
meaning he discerns. The figural significance of the 'three trees on the
low sky' ('Journey of the Magi,' l. 24), which are types of the crosses on
which Christ and the two thieves were crucified, or of the 'Six hands at

an open door dicing for pieces of silver' (l. 27), escapes his understanding. Since there 'was no information' and their destination was merely 'satisfactory' (ll. 29, 31), the wise man seems to concur with the judgment of Eliot's speaker in *Four Quartets*, who concludes with some exasperation that 'We had the experience but missed the meaning' ('The Dry Salvages,' 2.45).

Afraid that the recollected meaning will elude him and that he will be left with information devoid of knowledge, or with knowledge bereft of wisdom, the speaker is as desperate to commit his discovery to speech as the lover in Browning's 'Two in the Campagna,' who pleads with the auditor to help him hold fast the tantalizing thought he has just touched.

> And I would do it again, but set down
> This set down
> This:
>> 'Journey of the Magi,' ll. 33–5

The sudden shift into an imperative mood agitates the speaker. The deictics and anaphora of his repeated exhortations would sound less frantic if he were more confident of fathoming the mystery of his recollected experience and of communicating its meaning.

Most startling is the speaker's odd conflation of Birth and Death in his concentrated attempt to stabilize meaning by putting it into words:

> were we led all that way for
> Birth or Death? There was a Birth, certainly,
> We had evidence and no doubt. I had seen birth
>   and death,
> But had thought they were different;
>> 'Journey of the Magi,' ll. 35–8

His questioning recalls Thomas Becket's Christmas sermon on the Crucifixion in Eliot's play *Murder in the Cathedral*, where beginnings and ends, the Nativity and the Passion, are hard to keep separate. This birth was 'like Death, our death,' because, although it was a moment in time, it was also a moment out of time. It was a place of intersection between two kingdoms, the secular kingdom of knowledge and information that the wise men ruled and the prophetic kingdom of God, a kingdom of wisdom and fulfilment.

In his only use of a conditional, future-directed verb – 'I should be glad of another death' – the wise man, no longer at ease in the old dispensation, yearns for an experience that will take him out of time. Unlike the death that marks 'The very dead of winter,' this death reveals the unconscious wisdom behind the scholar's knowledge, the hidden antitypes behind the types. In the 'tavern with vine-leaves over the lintel,' the Death that clarifies the difference between mere memory and meaning-creating recollection, between 'information' and figural ciphering, lays bare the grammar or codebook of an apocalyptic symbolism that associates Christ with the vineyard, the temple, and the sheepfold in *Revelation*. Behind the hands dicing for pieces of silver the new figural way of reading deciphers a chain of biblical associations that include Judas's payment of thirty pieces of silver for betraying Christ and the soldiers' casting lots for Christ's garments. Like a scholar whose grasp of a paradigm transforms his understanding, the wise man begins to see how information relates to knowledge and folly to wisdom. Since 'Birth was / Hard and bitter for us, like Death,' it is even possible that Death itself, as Birth's mirror image, may be merely the last of the wise man's new beginnings. But this is a mystery he only partly pieces together and still only imperfectly apprehends.

Like the pilgrimage of Eliot's bemused Wise Men, the search for wisdom, hidden knowledge, or a fulfilling vocation is too important to be wholly voluntary. Even when the right choice is made, fulfilment may have to come in ways that are more spontaneous and less premeditated than we originally foresee. A diverting example is offered by Harvard's president Derek Bok, whose hope of being intellectually satisfied is not immediately realized when he decides to teach rather than practise law. Instead of 'mingling with scholars from a wide variety of fields' (1986, 1), Bok finds himself 'cut off' as a law professor 'from the rest of the University as if by a vast moat.' A 'bridge to Harvard Yard' comes for Bok in the form of an invitation from a young historian to play weekly poker games in a smoke-filled basement in one of the undergraduate residences. 'After two disastrous sessions,' however, Bok 'reluctantly conclude[s] that [he] could no longer afford to communicate with the rest of the University via the gambling table' (1986, 1–2). The table is not a true bridge, which must be constructed on piers of chance encounter and free association over longer spans of time.

Like our motives for choosing a vocation, a mentor's most important influences may also be unconscious. Dwight Culler is struck by the seeming contradiction of J.A. Froude's moving tribute to J.H. Newman. On the one hand, Froude praises Newman as the tutor who most deeply affected undergraduates and who 'seemed to be addressing the most secret consciousness of each of us.' On the other hand, Froude complains that Newman was 'not connected with a single effort to improve the teaching at Oxford or to mend its manners' (Culler, 119). Perhaps Froude's telling use of 'secret' provides the clue. The apparent 'paradox can be resolved,' Culler thinks, 'only by supposing that Newman's method was so informal that often people did not realize, when they sat under him, that this was the moment when their education occurred' (119).

Though no transcripts of Newman's tutorials survive, his Oxford sermons have come down to us. They are remarkable for their self-criticism and honesty. They also display the sympathetic imagination of a great scholar, who knows his adversary's position almost as well as he knows his own. Indeed, unless we read Newman's sermons with care we may mistake his own position for the enemy's. In sermon 15, for example, where the homilist is trying to explain how the development of true doctrine is possible, his sceptic seems to claim at once too little and too much for language. On the one hand, to speak of God at all is to take his name in vain. On the other hand, 'our ideas of Divine things are just co-extensive with the figures by which we express them, neither more nor less, and without them are not' (1887, 338–9). To apply an implicit empirical theory of language rigorously is to reach sceptical conclusions Newman wants to reject. But his Oxford sermons keep wresting truth from experience by demolishing the enemy from inside. Even when Newman tries to refute the sceptic by moving from nature to grace, traces of the sceptical argument remain intact. Any attempt to represent supernatural or eternal principles in intelligible signs must compromise, he concedes, the mystery of the subject. Like Johnson's poem 'The Vanity of Human Wishes,' Newman's sermons move Trojan horses into the enemy's citadel, where troops can be released for night battle. But victories are precarious and not easily won. Like Johnson, Newman arrives at truth only because he finds himself forced into ever darker corners of a world where phantoms betray each wanderer in the mist. The solution each apologist finds is the last refuge of a sceptical and critical mind in retreat from a series of alternative positions, successively found untenable.

Students of higher learning need theories about what to theorize and what to leave unexplained or unconscious. The importance of unconscious influence may help explain why Matthew Arnold and Froude both felt they were educated by Newman without knowing it. From Newman Arnold absorbed 'habits, methods, ruling ideas' (quoted by Culler, 119) that he was not conscious of assimilating at the time. Though Newman lived or embodied the truths that Arnold came to value most in later life, he never articulated them as a set of precepts or rules. Just as Newman never indoctrinated his students in Tractarian theology, so my own best instructor in philosophy, Emil Fackenheim, a former rabbi and authority on Judaism, never mentioned the Holocaust in his lectures. One feels that Arnold would never have become so indebted to Newman if the mentor's values and influence had not been hidden or subliminal. To become essential to everything Arnold did and said, Newman's most abiding legacy had to be unconscious.

It can be said of wisdom as it can be said of happiness or the desire to be a saint: we cannot attain a condition to which we too self-consciously or actively aspire. As Noel Annan notes, 'you cannot teach' wisdom, 'any more than to learn theology will give you a guarantee to be holy, or ethics to be good, or sociology an infallible blueprint by which to reorganize society' (1971, 14). J.S. Mill discovers that whenever he makes happiness his goal, it eludes him. And Tennyson's St Simeon Stylites and Eliot's Thomas Becket both discover that the desire to be a saint is inconsistent with the attainment of that state. There is a great line in Blake about 'a Moment in each day that Satan cannot find.' Perhaps the moments Satan cannot find will come to disciples of an unconditioned or unrehearsed life, as it comes to the lilies of the field in Christ's great sermon. The true end or goal of higher learning is not the mastery of ancient Greek per se or the intricacies of Hegel's logic. The point of such activities is that they have a goal which is different from their apparent goal. This is why Nietzsche says 'the spirit enjoys the multiplicity and craftiness of its masks' (160). He believes 'every philosophy also *conceals* a philosophy; every opinion is also a hideout, every word also a mask' (229). In pursuit of their most important and elusive goals, scholars and scientists come to live truths they cannot express. They embody values they can never know.

Just as essential as irony and intellectual midwifery to Socratic education is the neglected Socratic legacy of eros and enlightened ignorance in modern culture. In the *Symposium* Diotima explains that the god Eros is the offspring of Poverty and Resource. As such, Eros is his own opposite: he is both abundance and poverty. The word for poverty is the same as the word for the absence of resource, *aporia*. Knowledge that is at once an abundance and an absence is what Nicholas of Cusa calls *docta ignorantia* or learned ignorance. Paradoxically, the sum of such Socratic learning can be depleted or even exhausted at the same time it is replenished and increased. Like love, the more we squander a passion for scholarship or science the more it grows and is renewed. Unfortunately, learning also resembles religion: many scholars have enough of it to hate their rivals but not enough to love them.

The most valuable knowledge is often elusive as well as hidden. Not only do pockets of mystery exist in most disciplines but (far from eliminating them) increased knowledge also creates new mysteries. The more we know, the more we know there *is* to know: 'Hills peep o'er hills, and Alps on Alps arise' (Pope, 'An Essay on Criticism,' 2.232). The trailblazer in science or scholarship sometimes experiences the vertigo of Tennyson's Ulysses, who follows 'knowledge like a sinking star / Beyond the utmost bound of human thought.' As the normal science of Newtonian physics yields to the revolutionary science of Einstein, no foundation in science proves rock-bottom or solid. As Flexner says, 'physics and chemistry ... will not stay "put"; they have an elusive way of slipping through the fingers of the investigator' (18). As soon as medical science finds a cure for tuberculosis or typhoid fever, life is lengthened and a crop of new diseases appears. Since many problems are due (not to ignorance) but to knowledge, breakthroughs in medicine may paradoxically increase rather than diminish the range of puzzles to be solved.

I argued in chapter 8 that the distinguishing feature of an academic model or paradigm is that it can be exhibited in the performance of a master scholar or scientist without being explicitly formulated as a rule or stated as a precept. It can be shown but not said. If an apprentice does not see what is unique about a mathematical proof or a philosophical argument, there is little a professor can do except offer another proof or argument. After students have studied a sequence of these proofs, they usually apprehend the pattern or law that is generating the proofs. But either one perceives this pattern all at once or one does not. This is why each discipline requires a special kind of intelligence. One philosopher, Bryan Magee, compares it to musical aptitude. Some of his most intelligent students at Oxford 'were tone deaf to philosophy, and uncomfortably aware that they were not taking in what was going on, while others, less intelligent than they, were getting a tune out of it and enjoying themselves in the process' (314).

As Robert Frost tells Sidney Cox, 'clash is all very well for coming lawyers, politicians and theologians. But I should think there must be a whole [persuasive] realm or plane above that – all sight and insight, perception, intuition, rapture.'³ Teaching such neglected masterpieces as Christina Rossetti's lyrics and George Borrow's *Lavengro*, Frost offers courses on 'verdicts' or 'judgments' designed to show that original thinking is too often confused with mere compulsory 'voting – taking sides on an issue [that students] had nothing to do with laying down' (Pritchard, 172). In poking fun at the academy by having his many honorary degrees made up into a quilt, Frost even tempers his deeply serious attitude to play with an overflow of playfully irreverent animus. Frost's mockery of teaching is a refusal to teach. Like the knowledge Socrates imparts, the truths Frost values have to be shown or lived rather than merely talked about or named.

Always welling up at learning's source are two irrepressible impulses, game and play. According to Frost in 'Two Tramps at Mudtime,' the work of education is 'play for mortal stakes.' As William H. Pritchard says, 'one of Frost's favorite words for poetry was "Play," and everything he did or wrote could be thought about in the light of that notion. One of his notebooks contains perhaps his deepest word on the subject: "Play no matter how deep has got to be so playful that the audience are left in doubt whether it is deep or shallow"' (xvi). To the extent a scientist's games of discovery and a scholar's free play of intelligence lie within the non-purposive range of human motives, there is little we can say about them and little we can do to determine whether our response to them is deep or shallow.

Such uncertainty and silence point to what Richard Lanham calls 'humanism's central paradox,' the fact that its paradigm of game and play defeats the scholarly intelligence almost successfully. Having 'arrived at a boundary condition of our conceptual universe' (Lanham, 1983, 133), we find there are no more words, only the silence which marks the amazed possession of all words. At such a moment we can merely contemplate scholarship or science from a distance, as Ruskin contemplates great works of art or as characters in a play by Chekhov break an interval of profound silence with a memory that pierces the heart and suddenly transforms us.

The difference between Socrates and the sophists is that Socrates knows that wisdom, like holiness and virtue, is a form of hidden knowledge. It cannot be learned in courses on philosophy or religion, because it is a product of suffering and memory and of what Jacques Maritain calls 'hard experiences in friendship and love.' What is 'most paradoxical,' as Maritain says, is that 'all this extra-educational sphere exerts on man an action which is more important in the achievement of his education than education itself' (176). The soul of higher learning is what Spinoza calls *scientia intuitiva* or the intelligence of love, which cannot be taught. When the power-seeking magic of spiritual cranks, scientologists, and other gnostics loses its potency, the Nameless of a thousand Names removes himself from every disease of will and intellect, from every idol and every prayer, that now obstruct the way to him. Though Tennyson dreads the sundering of 'faith and form' in a 'night of fear' (*In Memoriam*, 127.1–2), he willingly invokes a 'strange friend' who is 'loved deeplier' as he is 'darklier understood' (129.9–10). The most important knowledge is hidden, because its glory and reserve are one. And yet the seer's removal from wisdom is his worst fear. For even to be in heaven, says Thomas Merton, is to love, 'without possession,' like God, who cannot possess even himself without becoming finite (475–6). The most remarkable fact about a scholar's wisdom or a prophet's insight is that his gift is nobody's property, not even God's.

Though I have been arguing in this chapter that aphorisms are often a signature of scholarly truth or prophetic power, their reticence and reserve are sometimes a mere trick of style. In a world of Heraclitean flux, where all things give way and nothing remains, the hurried jottings of Tennyson's epitaphs in *In Memoriam* or of Pater's aphorisms about a 'drift of momentary acts of sight and passion' in his Conclusion to *The Renaissance* may be the writer's only stay against a panic-stricken sense of the undertow of oblivion, the cataract of change that spends to nothingness. But instead of shrinking the truth into a maxim, an aphorism may also concede diminished expectations or even defeat. F.H. Bradley fears that 'our live experiences, fixed in aphorisms, stiffen into cold blood. Our heart's blood, as we write it, turns to mere dull ink' (1930, aphorism 27). 'An aphorism is true' only 'when it has fixed the impression of a genuine experience' (1930, aphorism 41). When Pater uses aphorisms in *Plato and Platonism* to present a theory of beauty that has stripped away Plato's metaphysics, or when Wittgenstein uses aphorisms to explore an inner world from which the Cartesian starting point of an I and its experience of objects has been removed, the elisions may be the only conceivable way of presenting an authentic vision of the world. But by omitting more than they retain, the aphorisms may also produce an oddly eviscerated picture. William James claims that wisdom is knowing what to overlook, both in the sense of justly overseeing and tactfully omitting. Yet an aphorism that leaves half of life unread casts doubt on the aphoristic force of James's own wise saying about wisdom. If the truth is the whole, as Bradley believes, and '"one never tells more than half," and in the end perhaps one cannot' (1930, aphorism 44), then what happens to the vaunted illusion of knowledge in a book of aphorisms? Is their promise of depth only a thin veneer of surfaces all the way down?

In asking what makes a gifted scholar or an inspired teacher, these last two chapters have touched on a mystery but made no attempt to solve a puzzle. Socratic education remains a mystery because scholarship and teaching are an art, not a technique or a craft. An imperfect solution to a logical puzzle is not a solution at all. But regrettably, a bad scholar is still a scholar and an inferior teacher is still a teacher. Since scholarship and teaching embody habits and practice, we judge them as we judge the conduct of a poem or the performance of a stage play, not as we assess a proof in mathematics or science.

In an age of science, any scholar who is faithful to the legacy of Socrates must honour in wonder the limits of what can be known or said about higher learning. Though science is one of the glorious achievements of the human mind, its conclusions can go no deeper than its premises. Physicists must realize that their thinking is incurably circular and only as profound as its axioms. To explain space in terms of time and time in terms of space, or to define such entities as mass, light, and energy wholly in terms of each other, is to fashion a tautology. It is also to concede total ignorance of what these convertible terms might mean in a wider context. Whereas Newton seems to have successfully imposed the laws of his mechanical system on nature, the quite different laws of Einstein's universe seem to be a free creation of his mind, subjected (of course) to the great institutionalized criticisms of physical science itself. Unfortunately, such mathematical entities as the physicist's concepts of timeless time and unending space do not appear to connect in any meaningful ways with the tensed time and limited space in which scientists themselves have always lived. Nor can science analyse the self, which is a material body with one extraordinary property: the ability to know itself from inside.

We shall never know whether death is the last of this self's new beginnings or whether the self survives the decay of its material body as the total creation out of nature it has made. Socrates, like Kant, insists that such mysteries as God and the immortal self are unknowable possibilities. 'My soul,' says F.H. Bradley, 'appears to be no more than a geographical expression, and where to find myself in it I have no idea' (1930, aphorism 57). In sharp contrast, sophists lack all sense of the mystery that surrounds the few things they do know. They fail to see that

'all the things that are of greatest importance to us are unknowable –
above all whether after this life,' in Bryan Magee's words, 'we shall be
plunged into timeless oblivion or go on existing in some way that is
unconceptualizable by us until it happens' (157). We emerge from the
inane, as Carlyle says, haste stormily across the astonished earth, then
plunge again into the inane (243). But where do we come from and
where are we going? I am astounded that so few thinkers are baffled by
the irreducible mysteries that surround the self, which is the most prom-
inent but least understood puzzle in the universe. Scientists must come
to see that there is more in heaven and earth than is dreamt of in their
philosophies. And scholars, who too often presume to know the mind of
the World-Soul as well as the World-Soul knows itself, must cultivate like
Socrates a deeper knowledge of their ignorance.

11

From Ivory Tower to Babel:
The Secret of the Maze

The sophistry I am keenest to refute is our culture's widespread assumption that to be educated as a scholar or critic rather than as a computer scientist or technician is to be educated as a misfit. I realize that uncritical nostalgia for the values of the age in which one's own youth was passed is often a mark of senility. But I like to think that in a new Dark Age of anarchy, in which barbarians have reduced to rubble the loftiest towers of commerce, the holy city of art and culture may still be remembered fondly as a *patria*. Higher learning may be fragile and unstable. The ivory towers may tilt and waver: they may even need structural reinforcement like the leaning tower of Pisa. But before we allow intellectual terrorists to raze the towers to ground zero we should know why our ancestors raised them and how they may still provide a homeland for the exile who is seeking beyond technology and consumerism – and beyond Babel's confusion of tongues – a community where scholarship and civility, science and enlightenment, may peacefully coexist.

An inquirer who seeks a single defining feature of higher learning – an attribute common to the maps and models of literature, history, philosophy, science, and professional education – seems at times to be lost in a maze without a plan. Sooner or later he experiences Yeats's alarm in 'The Second Coming': 'Things fall apart; the centre cannot hold; / Mere anarchy is loosed upon the world' (ll. 3–4). If pressed, however, to identify the secret of the maze, I should say that the Muse of higher learning – the scholar's master passion and the lifeblood of this monograph – is 'personal knowledge,' a phrase Michael Polanyi applies to the physical sciences but which I extend to every form of scholarship. A professor touched by the divine spark, who dwells on Parnassus with the Muses, appropriates knowledge as a living possession. As disinterested in principle as the objective knowledge of the new Prometheus, who is a robotic engineer called Frankenstein, 'personal knowledge' is the prize acquisition of both the old Prometheus and Spiderman. Though the Muse of higher learning cannot be cloned, she allows the prophet, rebel, and poet to soar, and the scholiast to weave intricate threadworks of commentary and gloss. Her energies of mind are the product of reflective inquiries which include the inquirer or view-holder as an essential feature of what is learned.

If aphorism is the preferred genre of prophets, the favoured trope of scholars is selected quotation. Since each quotation in this book once excited or moved me, it is my special way of being concrete and personal. But I admit that personal knowledge is more than just the memory of a mind set loose in the archives. To appropriate knowledge as a living possession is to locate it in a specific genre of discourse. Historical, philosophical, and critical essays comprise a distinct literary genre. Even though each essay uses a map or model unique to a specific discipline (as I try to show in chapters 4 and 8), they all claim a common ancestor in the genre of personal essay-writing invented in the sixteenth century by Michel de Montaigne. His talent for self-knowledge and for education of the whole person is still the informing spirit of a scholar's adventures in humane learning.

Since scholars are often rigorous logical thinkers whose first path to truth is rational inquiry, I have devoted chapters 3 to 6 to an anatomy of scientific, scholarly, and practical knowledge in our culture. Though all knowledge, including scientific knowledge, is personal, its power to reflect or embody the view-holder's values is minimal in technology and at a maximum in idealist philosophy and the arts. A true scholar realizes that a science that excludes the view-holder cannot suddenly restore him to its picture of the world. For this reason any form of knowledge designed to substantiate the claim that the scholar is by birth and by nature a prophet and a poet, a self-conscious fashioner of worlds, must include the interpreter himself as an important part of what is studied.

My theory of personal knowledge rejects the positivist's claim that knowing makes no difference to what is known. Let us assume that the mighty world of eye and ear consists of two parts, 0 and c: the world an observer knows, c; and the world that is known apart from any observation of it, 0. According to the positivist, 0+c is the same as 0. We are asked to believe that the glory of the world, that warm and breathing beauty of the flesh the heart finds delightful, is identical with a ghostly abstraction of it. There are two difficulties with this positivist anxiom. In the first place, not even a physicist can directly observe the unearthly ballet that the atoms and subatomic particles of quantum mechanics are said to perform. A scientist can posit these phantoms only with the aid of precision instruments and highly sophisticated experiments. In the second place, positivism involves a logical error. As Collingwood explains, 'if you know that no difference is made to a thing 0 by the presence or absence of a certain condition c, you know what 0 is like with c, and also what 0 is like without c, and on comparing the two find no difference. This involves knowing what 0 is like without c; in the present case, knowing what you defined as the unknown' (44). A positivist who performs the illogical operation of claiming to know a world that he removes from every act of knowing it is trying, ironically, to know *less*, not *more*, than can be known. If the positivist's knowledge of an unknown world is as true as he claims it is, then it must also be false.

I may approve of Mozart and Tennyson because I love what they create. But I cannot love Edison or the Wright brothers because I approve of the inventions they discover. A humanist's love is the passionate attempt to find what is unique to himself, not in a light bulb or an airplane, but in another person. Though there is always an observer in the picture, his influence is less pervasive in an anatomy of knowledge than in a testament or a confession. But even in a confession by J.H. Newman or Benvenuto Cellini, the autobiographer can be understood for one of two opposite reasons: either because he always tells the truth or because he never does. Since people, unlike computers, are seldom so consistent, I now believe that the personal knowledge unique to each autobiographer and scholar is an undiscovered country. An adventurous traveller who is daring enough to approach its border may leave behind a hint half guessed. But no Lazarus has come back to tell his story.

Using technology to discipline the Muse, a funding agency is less likely to commission a symphony by Mozart or a dialogue by Nietzsche than a new computer study by H.T. Buckle on how the Corn Laws' repeal affects Victorian birth rates. A large research team is always more in fashion than a wayward philosopher who might write another *Republic* or a prophet who might preach a new Sermon on the Mount. A technological culture woos the mind with computer software, not with the eros of persuasive speech. But as soon as the brain is wired to digital technology, higher learning becomes a battle against the bewitchment of the mind, not by means of seductive language, as Wittgenstein fears, but by an assault of information theory, computer screens, and a stream of unselective data on the Internet. Once 'software becomes internalized and consciousness [has] to grow a second skin,' George Steiner wonders how long 'commandments of love and rebellion,' 'the romance of the persona in the pedagogic act,' will be able to survive (180–1).

When I started teaching in universities more than forty years ago, the tradition of membership in a venerable community of learned men and women was still possible to honour and maintain. Colleges and universities were not yet run by bureaucrats who saw faculty members as their adversaries, or by faculty unions eager to fight for some real or imaginary right. To rebuild the ruins ill concealed behind their trailing ivy and crumbling towers, universities have to repudiate what Huxley calls 'the gladiatorial theory of existence.' To this end they should abolish chairs in how to be competitive, scrape away such costly accretions as athletic stadiums and bread and circuses for the alumni, and repress the self-assertion of media stars and gurus. Renewing their love affair with learning, universities must also include more of the nation's gifted poor. When positivists are not betraying the scholars by spreading the heresy that all learning aspires to the condition of a physical science, entertainers are too often selling them out by assimilating Nietzsche's dark genius to rap or heavy metal, and the scathing prophecies of Blake to the lyrics of Bob Dylan. Bewitched by the commercial success of a circus or a Hollywood movie set, which dictate their values and feed their dreams, universities are in danger of substituting easily marketed facsimiles of education – big lectures by superstar professors – for a renewed commitment to learning. By bending Titans of higher finance to their will, administrators should try instead to fund and honour unglamorous scholars and scientists, the wizards of higher learning, and do everything in their power to moderate the obscene credentials game that rewards the pushiest extroverts and highest bidders.

Bill Readings believes that scholars should be suspicious of recent attempts to replace Kadish's educational maps and models or Oakeshott's civilized 'conversation of mankind' with what Readings calls 'a discourse of excellence' modelled on the ethos of the business corporation. An 'ideology of excellence' has as such no content. 'What gets taught or researched matters less than the fact that it be excellently taught or researched' (13). One legacy of 'excellence' is the growth of cultural studies and their flight from traditional models of literary, historical, and philosophical learning. Readings concludes that 'what is crucial about terms like "culture" and "excellence" (and even "University" at times) is that they no longer have specific referents' (17). In such a university 'a general principle of administration replaces the dialectic of teaching and research, so that teaching and research, as aspects of professional life, are subsumed under administration' (Readings, 125). When technology and the 'ideology of excellence' create a sick or insane environment, cultural neurosis may set in. In such an environment sophists who value the quantitative forces of bigness – big enrolments, big lectures, big departments, and big budgets – need to be reminded that the modern Socrates is often a philosopher writing alone in a library or a chemist working with a few dedicated colleagues in a laboratory. An obscure refuge is precisely the place education is most likely to occur whenever an original thinker and a few select students gather together to launch an inquiry or explore a new problem. The university must honour the energy of these solitary talents and deplore the inertia of commonplace minds. Henry Adams claims that 'any large body of students stifles the student. No one can instruct more than half a dozen students at once' (302), since discoveries and insights are never mass produced. While the large professional and business schools are always on fashionable first-floor display, the genuine scholar is often confined to a dusty closet in the attic of the campus's Gothic asylum, like the madwoman in *Jane Eyre*.

A university's symbolism is usually associated with its architecture and geography. The symbolic centre of the University of Toronto is King's College Circle, where the university library, the press, and the departments of French and English used to be quartered. To make more room on the Circle for the massive fortress of the ever-expanding medical and engineering schools, the library was moved three decades ago to St George Street. When the French department's home was swallowed up by the sprawling medical library, the department was exiled to a new wing of St Basil's Church on the eastern edge of the campus. From the seclusion of St Michael's College the French and German departments can now view the soaring towers of Toronto's financial district on Bay Street. Evicted from its quarters on the prestigious Circle, even the University Press has had to adjust to the seventh floor of an office building over a donut shop and food court on Toronto's busiest commercial street. The ritual was about to be consummated with the banishment of the English department, the last humanist holdout on venerable King's College Circle, to a decaying Charles Adams mansion on the western periphery, replete with Gothic gables and the ghost of a recently murdered professor of fine art. But the Gothic mansion proved too cluttered and antiquated. Instead, the transformation of Canada's largest university into a polytechnic in all but name was completed when the medical building gobbled up the English department, which found itself like Jonah in the belly of the whale that had recently tried to spew it out.

The sad fact is that many schools still promote the myth that if young people attend the right colleges, meet the right people, or join the right church or political party, they will be prosperous, respected citizens. But as Alan Ryan says, democracies support universities because they 'think the cultural and intellectual achievements of the human race are the inheritance of all of us, and not only of a favored few' (21). Unfortunately, many a mute inglorious Socrates drinks hemlock too soon. For 'death is by no means always a mere cessation of heart and brain activity' (Price, 66). As one professor notes, 'anyone who's taught college, as I have for four decades, well knows that a number of people choose life-long mental and spiritual death in late adolescence, if not sooner' (66). When a member of my own graduate seminar kept waving his hand last week, I was at first surprised by his subdued look when I asked face-tiously, 'What's on your mind, if you have a mind?' Obviously, *I* was the mindless one. In trying to be funny while venting my secret rage against the early death of mental curiosity, all I did was hurt him. I hope I won't hurt people. But I know I can't be counted on to do generous and cour-teous things. When I meet a scholar on the brink of mind-death, whose last flicker of intellectual hope or ambition is guttering out in front of me, I am never sure there will come a flash of the right thing to do.

The dispiriting truth is that the vast majority of students and many colleagues are non-scholars. The world is too much with the scholar: acquiring knowledge and ideas, then spending that capital in seminars he gives and in books he writes, sooner or later the scholar lays waste his powers. Without the fellowship of a few close friends who can share ideas, the professor subdues himself to what he works in, like the dyer's hand. Today's universities are run more and more as business corpora-tions by fiercely competitive CEOs who have ceased to value the secluded life of prolonged concentration and freedom from anxiety that productive learning requires. Since most scholarship is an art not amenable to mass production methods, it is a wonder that more human-ists are not 'as terribly oppressed by the present tensions' as the Harvard scholar F.O. Matthiessen, who writes in a final suicide note: 'I can no longer believe that I can continue to be of use to my profession and my friends' (Bate et al., 99). Day after day and year after year, exposing his soul to non-scholars with no passion for the life of the mind, the scholar may feel less like a lover of wisdom than a prostitute or a whore.

It is sad that scholars should often feel bitter and betrayed. But perhaps this is how modern disciples of Socrates ought to feel. Pre-college education consolidates social authority by transmitting a knowledge of its ideologies and myths. But in turning each prejudice inside out, the models of higher learning also subvert that authority. At the heart of higher education is an aspiration toward freedom and play for their own sake, unmotivated by prudence or foresight. The scholar or scientist pursues an imaginative and intellectual vocation, not because it pays well or allows him to live a life of diligent idleness, but because like writing a poem or proving a theorem it is a satisfying and beautiful thing to do. By revolutionizing its models, higher learning fulfils Socrates' high vocation of corrupting the youth and of challenging the judges who would sentence them to life imprisonment in Plato's cave.

Observers with no experience of scholarship or teaching wonder why passionate thinkers in the academy often burn out. It is no mystery to the scholars. To be an academic celebrity it is often more important to be provocatively wrong than creatively right. Even when a scholar like Edward Said tries to transmit an original but just interpretation of Orientalism to posterity, he has to walk a tightrope: the model he proposes must be neither too rigid nor too fluid. If the model is too yielding, it will be displaced by unworthy rivals and retain no discernible identity. If it is too rigid, no intelligent opponent will keep it alive by feeling free to challenge and change it. A tradition in scholarship or art, if it hopes to survive, must have adversaries worthy of attacking or even trying to destroy it, and so of enlarging or contracting what it has to transmit.

As Bacon says, 'in this theatre of man's life it is reserved only for God and Angels to be lookers-on.' Dulled by routine, the scholar who has wasted too much of his life supervising theses and serving on committees may decide to spend his retirement 'in studies and contemplations, without looking back.' If a professor has any regret, it is for years misspent on academic politics and petty administration, on 'things for which [he] was least fit; so as [he] may truly say, [his] soul hath been a stranger in the course of [his] pilgrimage' (Bacon, 542). The highly technical, impersonal knowledge that professors of dental surgery share with their students is only a small portion of the surgeon's total endowment, intellectual and human. Though Ben Jonson can proudly say that his best piece of poetry is his dead son, no dentist would boast on his gravestone that he is filling his last cavity. Literary scholars and philosophers are more like poets than dentists: they give away their marrow and essence. In their scholarship and teaching they offer themselves. The important but elusive bond between scholars and their mentors is not entirely intellectual but includes friendship and love. Scholars share what friends and lovers share, a part of themselves.

The university must save its own saving remnant – its cadre of dedicated scholars, scientists, and apprentices – from the suburban middle-class consumers whose sole utopia is the felicity of the defining sports car or social club that identifies for all time who they are. As the church, mosque or temple of secular culture, the university wakes the mind to maps and models it had not yet imagined. But to a slumbering dogmatist any wake-up call may be unwelcome. No consumer in a supermarket wants to buy an imitation grapefruit when a real one is available. But many consumers of higher learning prefer an imitation education to an education that may interfere with their career plans by transforming their values or changing their lives. A university may therefore need to challenge and even disappoint consumers who want less for their money than a university needs to give them. The well-kept secret of higher learning is its protection *against* rather than *for* the professions. But like Leo Strauss's readings of Plato or Moses' first set of Tables, this doctrine is too dangerous to pass on. A university must protect its secret by sustaining its own protective fiction about being directly useful to society. Even in perpetuating its noble lies, however, the university must be true to its own subversive function. It must 'blow' its students' minds, not with drugs, but ideas.[1]

The Socrates whom humanists admire is not the Platonic inventor of dualisms between knowledge and mere opinion, or between dialectic and rhetoric, but the genius responsible for what Richard Rorty calls 'the very complex, shifting, dubiously consistent thoughts' of the dialogues (1999, xii). Humanists in the university should provoke students to be self-creative. Instead of repeating St Paul's dictum that the truth will make them free, they should seek free inquiry as an end in itself. For even if truth is delayed, freedom itself should not have to wait. Richard Rorty believes that 'the only point in having real live professors around instead of just computer terminals, videotapes and mimeographed lecture notes is that students need to have freedom enacted before their eyes by actual human beings.' He concludes that such 'enactments of freedom' are 'what non-vocational higher education is all about' (1999, 125).

But freedom is a product of discipline, and cannot be taught by a professor whose prior aim is to subvert authority. Rorty assumes that pre-college education can socialize its students and that universities can

then educate them to be free. But the two activities are surely as inseparable as Socratic scepticism and the ideas that Socrates criticizes. Rorty is
right to emphasize that non-vocational higher learning is revolutionary
and subversive. But except in his essay 'The Unpatriotic Academy,'
where he attacks left-wing professors for isolating themselves from a
society they should patriotically criticize and try to reform, Rorty fails to
see that education is also a deeply conservative enterprise.

To the extent that it must oppose a culture's repressive authority and
its pressure to conform, higher learning is liberal and even subversive.
But it is also radically conservative, because it is always trying to subvert
authority to renew some traditional allegiance, some commitment to an
academic discipline, a culture, or what Frye has called a 'mythology of
concern.' Universities must accept Rorty's challenge to invent 'new
forms of human freedom' by 'taking liberties never taken before' (1999,
126). But they should also remember that liberal values are best communicated indirectly, not through political indoctrination but through
some analogy with the student's mastery and transformation of an academic discipline like history or physics. Any education in metaphor is
also an education in freedom, since metaphor itself, the most subversive
rhetorical trope, offers a direct challenge to semantic authority by dis-
mantling and reinventing language in phrases like 'information highway' or 'surfing the Internet.'

Rorty is right, I think, to reject Plato's distinction between truth-
seeking disciplines like dialectic or logic and opinion-forming disciplines like rhetoric. Nothing could be more misleading than to say that
science gives us truth, philosophy meaning, and poetry mere pleasure.
As Kant realizes, there are different kinds of truth. Conceptual understanding is a product of disciplines like chemistry and physics which
organize our sense perceptions of the natural world. Ethics, by contrast,
is a construction of the practical reason designed to provide indirect
access to such truths of the Pure Reason as God and the immortality of
the soul. The artist's intuitions belong to Kant's third realm of pure purpose, or purpose devoid of practical utility. No inquiry is cognitively
superior to any other inquiry. Each has virtues that are as unique as the
distinctive qualities of roast beef and sherbet.

In the chapter on personal knowledge I tried to show how Thomas
Kuhn and Michael Polanyi complete Kant's revision of Plato. Instead of
differentiating among disciplines on the basis of how they know and
their claim to be 'true,' disciplines are judged according to the stability
or instability of their paradigms. As Rorty says, 'reading Kuhn led me,

and many others, to think that instead of mapping culture on to a epistemo-ontological hierarchy topped by the logical, objective and scientific, and bottoming out in the rhetorical, subjective and unscientific, we should instead map culture on to a sociological spectrum ranging from the chaotic left, where criteria are constantly changing, to the smug right, where they are, at least for the moment, fixed' (1999, 180). Rorty implies that thinkers display most intellectual freedom at the left end of the spectrum where academic disciplines are in their most revolutionary phase. But I think this assumption is wrong. Great breakthroughs can seldom be made at either end of the spectrum. Only when a shift from one zone to another is about to occur does revolutionary science or scholarship become possible for a limited period of time.

Newton could not discover the laws of motion during the Middle Ages when Aristotle's physics formed too stable a paradigm. But neither could a great hedgehog like Hegel appear in the twentieth century, when philosophy's models were so unstable that there was no international consensus about whether analytic or continental philosophy was more reputable. Spun round today like a frenzied dervish, the modern scholar shares Bacon's fear that a 'promiscuous liberty of search' has 'relaxed the severity of inquiry' (432). Though opportunities exist for achieving distinction (or possibly even greatness) in every discipline, the bold discoveries that leave future generations breathless occur only at rare moments of transition. They take place when the stable models of late medieval culture are about to yield in Shakespeare to less stable Renaissance models, or when the instability of physics at the end of the nineteenth century is about to give way to the stabilizing synthesis of Einstein.

Despite what Rorty says about professors of dissent who are hired for the express purpose of making students self-critical and sceptical, such values (however admirable) are always a function of trying to think and experiment with the aid of specific models, especially with unstable ones. Scepticism and the freedom to dissent can no more be instilled in an intellectual void, in an absence of protocols and paradigms, than faith can be inculcated in the absence of some stable body of doctrine that theologians are free to affirm, deny or amend. One can never function effectively as an example of what Frost calls a 'symbolic' as opposed to a 'real' teacher or as a self-appointed minister of disturbance in general.

The ideal university is a city of the mind that exists only *in* the mind. But whenever it tries to find a local habitation and a name, the university must cultivate a tolerance for paradox that allows it to dwell in two worlds simultaneously. As a centre of scholarship and pure research, the university will be most useful to society by refusing to be useful. It will achieve its purpose best if it disavows purpose in favour of the free play of a scholar's contemplative inquiries or the more competitive, heuristic games of science. The scholar is a purist, an ascetic with as much distaste for utility as Hawthorne's 'artist of the beautiful.' In questioning convention, he will both respect and criticize his culture. He will simultaneously honour and reject tradition.

I argued in chapters 9 and 10 that the scholar's most important intellectual habits must be practised rather than professed. The literary scholar or historian displays or exhibits his tenacity, elegance, tact, or wit. But he cannot talk about these virtues without profaning them. Though Socratic wisdom may seem redundant to sophists seeking quick entry into well-paying professions like medicine or law, the innovative scholar or scientist is always the opposite of redundant. Despite Albert Jay Nock's disclaimer, the autobiography of a scholar is never the memoir of a 'superfluous man.' For in mechanically replicating a model, it is the hack or technician, not the trailblazer like Socrates or Jesus, who is expendable. Individual contributions to culture are never superfluous in the way acquired characteristics in an individual animal are redundant to the species. Only in a totalitarian society is the critic of authority superfluous. The thinking minority is redundant only in a slave culture willing to submit to the will of a mindless majority.

Socrates is to scholarship what Prometheus is to technology and science: its hero and deliverer. Just as Faust, the would-be god who sells his soul to the devil, is a fallen Prometheus, so the sophist is a caricature of Socrates. Whereas the so-called expert or authority markets wisdom by presuming to think for others, Socrates acquires wisdom by thinking for himself. Socrates' most dangerous enemy is the efficiency expert who uncritically assumes that scholars and scientists cloistered in their archives and laboratories live apart from the 'real world,' which is said to be the world of money managers, lawyers, and civil engineers. I challenge their assumption by arguing that the so-called 'real world' is a delusive version of Plato's cave, a shadow world of idols and illusion. Technology is not science – as commerce is not culture and grammar not literature. Instead of consuming their lives in chasing phantoms, scholars who contemplate Homer or the geometry of non-Euclidian space dwell in a world of substance rather than shadow.

In an age of technology and genetic engineering it is salutary to recall Walter Pater's *Plato and Platonism*, which claims that, like the art of the Platonic dialogue, the highest knowledge for Socrates is knowledge of a single, living person. The 'real world' is *inside* the university. It is not outside the academy, but at its heart or centre, wherever the living energy of the individual scholar or scientist allows him to contemplate his models or test and redesign his paradigms. Whereas Prometheus is a benefactor who energizes science, Socrates is a wise man whose dedication to thought animates personal knowledge and humanizes scholarship. The sophist, by contrast, is a wise guy, a facile 'know-it-all' or sly casuist with quick 'fixes' for everything. Redressing both the foolish cleverness of the sophist and the clever folly of the media guru is the wise folly of Socrates. Instead of applying rules inflexibly, Socrates possesses the 'negative capability' Keats associates with Shakespeare, an ability to remain 'in uncertainties, mysteries, doubts, without any irritable reaching after fact and reason' (Keats, 261). Great scholars are distinguished from pedants, not by their intellectual ambition or their tolerance for brain-numbing labour, but by the wager they stake on personal knowledge and by their enormous capacity to grow.

Since every scholar needs a Socrates and every student needs a mentor, Newman and James Cameron both stress the importance of conversation and personal influence in higher learning. Almost every commentator on education from Socrates to Harold Bloom has praised the power of an erotic bond between teachers and their students. I suspect that Bloom's fascination with 'the educational genius' of Shakespeare's Falstaff, whom he calls 'the Socrates of Eastcheap,' has something to do with a mentor's love for a student who betrays him. 'Falstaff dies for love,' Bloom reminds us, 'a teacher's love, I would emphasize' (2002, 23–4).[2] Though Richard Rorty concedes that 'love is notoriously untheorizable,' he believes that 'erotic relationships are occasions of growth, and their occurrence and their development are as unpredictable as growth itself. Yet nothing important happens in non-vocational higher education without them. Most of these relationships are with the dead teachers who wrote the books the students are assigned, but some will be with the live teachers who are giving the lectures' (1999, 126).

Philip Marchand recounts the extraordinary conversations that Marshall McLuhan used to have with his graduate student Hugh Kenner, which often lasted ten hours at a time. Even Northrop Frye cherished his long Monday afternoon sessions on theology and literature with his most brilliant student, Peter Fisher, who told Frye he should learn more

Sanskrit if he wanted to understand Blake. Perhaps Thomas Merton comes closest to explaining the importance of such intellectual friendships. 'In every letter,' Merton reflects, Dom Porion 'comes out with some phrase or sentence that so expresses what I love or think or feel, or what I desire, that it is as if my own soul had spoken within me' (351). We welcome intellectual as well as spiritual friends because 'seeing your own thought objectified in the mirror of another makes you return with greater profit to your own mirror. Such correspondence, far from being a distraction, is a huge grace' (Merton, 352).

Like the Cambridge undergraduate in E.M. Forster's novel *The Longest Journey*, all scholars yearn for 'a society, a kind of friendship office, where the marriage of true minds could be registered.' We know that nothing can irradiate a subject like the saving imagination of a close friend or colleague. But the irony of intellectual friendships is that they are at once fragile and strong. For a few years at college friends come together, 'like straws in an eddy,' only 'to part in the open stream' (Forster, 69). My best companion and critic is my wife, a scholar in her own right, who has given me the precious legacy of children. I know I am not entitled to her gifts, and sometimes I think a scholar's life should be non-conjugal. Henry Adams says 'one friend in a lifetime is much; two are many; three are hardly possible. Friendship needs a certain parallelism of life, a community of thought, a rivalry of aim' (312).

Though I realize I am lucky to have had two or three companions who meet Adams's criteria, a shadow fell when one of the friends who most illuminated literature and religion for me entered medical school. I can still enjoy informed conversations with my daughters and my wife about physics, geology, and even animal psychology. But I can no more digest anatomy, physiology, and clinical medicine than I can eat lobster with a shell on. The human body's surreal dance of categories, however fascinating to a medical scientist, is no substitute for the philosophical or literary discussions my mind found delightful. Who can explain who or what we love? As Morley Callaghan has said, 'there may be, in the exchange of a glance, a recognition as profound as that old intense Greek recognition of a blood relationship or bondage' (quoted by Barry Callaghan, 558). For those who travel to the undiscovered country, who wager everything on personal knowledge, the loss of such bonds is irreplaceable. Like separation or death, it diminishes each of us.

But no single friend or mentor, and no mastery of a discipline, can save the scholar from what Ernst Cassirer calls 'the tragedy of culture.'

The city of art and culture may seem to some scholars and artists a holy and everlasting city. But we never truly possess Yeats's Byzantium. Its treasures remain untouchable, unviolable, because we are removed from art and culture at the precise moment we are closest to them. When we try to take possession of the books we write, the scholarship and science we produce, or even the friends and mentors we love, 'all that is sweet in them becomes bitter,' as Merton complains, 'all that is beautiful, ugly' (368).

To grow old, as Philip Larkin says in a touchingly honest poem called 'Afternoons,' is to feel something pushing us to the side of our own lives (ll. 23–4).

> Young mothers assemble
> At swing and sandpit
> Setting free their children.
> …
> Before them, the wind
> Is ruining their courting-places
>
> That are still courting-places
> (But the lovers are all in school),
> And their children, so intent on
> Finding more unripe acorns,
> Expect to be taken home.
> Their beauty has thickened.
> Something is pushing them
> To the side of their own lives.
> 'Afternoons,' ll. 6–8, 15–24

'In the hollows of afternoons' (l. 5), when the leaves have fallen in Queen's Park and the Arctic wind is beginning to batter the halls he once lectured in, a retired professor may feel less gently pushed than Larkin's mothers. A scholar who has been replaced by an unqualified rival or by no successor at all may feel that he and his life work have been trivialized. One of the saddest moments in Alvan Kernan's memoir is the time of his early retirement from Princeton, 'which seemed to feel no regret for the passing of the old order.' Instead of welcoming a younger generation of scholars as an older generation had once welcomed him, Kernan fears he has reached the end of a line. There was a 'desire,' he says, 'to get the fossils out of the way so that younger people,

women and minorities in particular, could be appointed in our places ...
There was no real antipathy, at least so I believe, between me and my
younger colleagues, but neither was there the sense of affinity, of a
shared, continuing search for truth, that had made me one with my own
elders in an earlier time' (1999, 293–4). Though it is disturbing to think
that 'the search for truth had largely been replaced by teaching and
publication as careerism and political action' (294), it is possible to
think of worse scenarios.

More pathological and malignant surely is the silent rage of Salieri,
the accomplished artist who has the misfortune to be a colleague (and
rival) of a rare and envied Mozart. If a professor teaches long enough he
may encounter a gifted student or colleague who has a daily beauty in
his life or a felicity in his thinking that reminds the older scholar of his
own failings. 'The secret of happiness is to admire' the achievement of a
great colleague or rival without desiring it for oneself. But as F.H. Bradley wryly adds, 'that is not happiness' (1930, aphorism 33). To come to
terms with genuine excellence in other scholars, it may be tempting to
deny that excellence exists. Yet in such acts of denial we not only help
destroy excellence: we also destroy ourselves. Goethe says there is only
one way to accept with grace the genius of another person: and that is to
love that person and his gift, as we love Mozart and his music.

To the Oedipal poet who murders rather than honours his father,
inheritance is a shackle to remove, not a gift to be grateful for. But
among such immortals as Homer and Virgil, Dante and Shakespeare,
parricide is as unthinkable as rivalry among the persons of the Trinity.
When Coleridge hears Wordsworth read *The Prelude*, his response is not
the testy grudgingness of Pope's Atticus, who bears, 'like the Turk, no
brother near the throne' ('The Epistle to Dr. Arbuthnot,' l. 198).
Instead of envying his friend, Coleridge is spontaneously moved to
quote from an earlier poem by Wordsworth, thankful for 'the human
heart by which we live.' Wordsworth had ended his great ode with the
reflection that to him 'the meanest flower that blows can give /
Thoughts that do often lie too deep for tears' ('Ode: Intimations of
Immortality,' ll. 201, 203–4). Subdued by 'thoughts ... too deep for
words' ('To William Wordsworth,' l. 11), Coleridge now expresses his
gratitude to Wordsworth by graciously alluding to an ode much greater
than the poem Coleridge is writing. When he rises in a mood of fearful
joy, he finds himself in prayer.

The people and books that matter most to scholars and that 'can be
loved for their own sake alone' come in the end to be 'invisible and

unimaginable and untouchable, beyond everything else that exists' (Merton, 368). Each scholar or mentor carries the grave with him. For when he tries to revive the joy of a shared idea, the intoxication of the first book he published, or his first love affair with learning, he can seldom renew its mystery. Even as he brings light to another mind, he also puts out that light unless he allows the most wayward qualities in each disciple to flourish and grow. The disciple-apprentice must freely become who he is, emerging out of himself and completing himself, even as Judas, the critic and betrayer of the master. In the holy city of Byzantium nothing that the scholar and artist 'can see or hear or touch will ever belong' to him. An absurd desire to give ourselves away to the beauty that 'eats out' our 'heart' and to possess the souls of scholars and mentors who ravish our minds, is, as Merton concedes, 'an unworthy desire.' But as Plato understood, it is also a desire few can avoid, since the soul of higher learning, the eros that urges the mind toward attainment, 'is in the hearts of us all' (Merton, 369).

When Yeats says that the intellect of man is forced to choose 'perfection of the life or of the work,' he is identifying a choice between learning that is liberal but not professional and learning that is professional but not liberal. A humanist may touch some numinous moment in Wordsworth's poetry or Thoreau's prose that transforms him. But even in experiencing a joy so pure that it is passionate, a contemplative scholar is doing nothing to promote social justice or to find a cure for cancer. And a research scientist who fights a losing war against disease is seldom able to experience the subdued elation of Kant's moral philosopher, who explores a realm more sublime, Kant thinks, than the heavens charted by astronomy. Nor is Keats's shock of discovery when first reading Homer commensurate with the saintly Alyosha's illumination in *The Brothers Karamazov* or with the bliss of Wordsworth's mystic, whose mind is a 'thousand times more beautiful than the earth' (*The Prelude*, 14. 449). All these goals 'seem desirable,' as Richard Rorty says, 'but I ⌊do⌋ not see how they could be fitted together' (1999, 10). If they converge at all, they meet only like a parabola and its asymptote at the end of time and the world, in a vision that absorbs all differences in itself.

Literature's special contribution to higher learning has something to do with the injustice we feel whenever our idiosyncrasies and passions, which we believe to be unique to ourselves, are described in the abstractions of the psychological sciences or ethics, which use concepts common to everyone. We know that, like a psychological treatise, *In Memoriam* speaks to and for others about our common humanity. But because a poem's representation is wholly individual, we may find in it an image of ourselves that is an invention in both senses of that word. Its meaning is something we come upon or discover, a lucky find, but also something we alone can make of this find. We are all defined by what we have lost. But since the mystery of a great elegy comes not from anything bewildering in the world but from something unlimited in the subject, *In Memoriam* can also speak across the ages, across the reader's death, about the deepest puzzles in his own life.

In a culture in which science occupies the same place in thought as did theology in the Middle Ages, how are scholars to justify their vocation? The value and function of scholarship are closely associated with the fortunes of a liberal education. The word 'liberal' has both a negative and a positive implication. Negatively, a liberal pursuit like the study of poetry is not engaged in for the sake of some immediate advantage, as is commerce or finance. Positively, a liberal pursuit is carried on for its own sake because it is enjoyed or preferred. T.H. Huxley's objection to a literary education, like Bentham's objection to poetry, uses the merely worthwhile nature of such pursuits to trivialize them. But despite Huxley's high-spirited banter against the 'monopolists of liberal education,' most humanists today, like their Victorian precursors, are too self-critical and playful to see themselves as high priests or 'Levites in charge of the ark of culture' (Huxley, 3:136). As a scholar and a poet, Matthew Arnold knows he is fighting a rearguard action. But he never wavers in his faith that poetry has healing power and an educative function. He is confident of Wordsworth's capacity to domesticate the strange and of Coleridge's ability to estrange the familiar. In seeing life steadily and seeing it whole, and in becoming a part of the best that has been thought and said in the world, Arnold's literary scholar can also take a disinterested, inclusive view. He knows that any profound encounter with the past partly cancels 'the doubtful doom of human kind' by linking the generations. In a civilized society a study of poetry and the traditions it honours can confidently be left to justify themselves. Like the pursuit of virtue or a liberal education, they are their own reward.

I have been drawn all my life to the nineteenth-century English poets, possibly because they have a talent for redressing the giant agony of the world by educating intelligence into spirit and exploring hiding places of power. Keats's letter on soul making and Wordsworth's Intimations Ode are sacred texts. Though Wordsworth is assured that Nature is a temple, a sanctum or holy place that 'never did betray / The heart that loved her' ('Tintern Abbey,' ll. 122–3), he also offers 'thanks and praise' for moments of fearful awe. Any serene intuition of benevolence at the heart of things is tempered by a mystical chill, a zero at the bone, that makes him tremble 'like a guilty Thing surprised' before what Rudolf Otto calls the *mysterium tremendum* – the ultimate mystery – which Wordsworth approaches with awe, fascination, and a sense of 'creaturehood' ('Ode: Intimations of Immortality,' ll. 141–8). Like Wordsworth's two poles of revelation, the kindred points of joy and awe, the absences and losses in Tennyson may ebb away in grief in the lines to Dufferin or mount to subdued elation in 'Tithonus' and 'To Virgil.' In both cases his grief is not just a shock or surprise but also a liberation.

Poets as different as Tennyson, Browning, and Hopkins each try to replace the senseless compulsion of a world 'red in tooth and claw' with that persuasive unity of power, love, and knowledge foreseen by David when he baptizes Browning's faculty psychology in the name of the Trinity at the climax of 'Saul.' Even in 'Dover Beach' the unbelieving Arnold counters the melancholy, long withdrawing roar of the sea of faith with an appeal to Sophocles, with medieval therapies for the agony and strife of human hearts, and with a memorable allusion to Wordsworth's power both to hear and subdue 'the still sad music of humanity.'

When Tennyson steps off the moving platform of 'the great world's altar-stairs' in *In Memoriam*, he is swallowed up in an abyss of grief and dissolution. He never seems to get over the impermanence of everything he sees, the fragility of his love for Hallam, the taste of death in the new geology of Chambers and Lyell. But as a poet of 'honest doubt' who suffers from Pascal's fear of terrifying waste spaces, Tennyson cannot bear the thought of life's being ultimately meaningless. The mourner in *In Memoriam* is so stunned and drained of energy, so paralysed by a sense of comprehensive wrongness, that, unlike most pastoral elegists, he seems not quite up to the task of writing a poem at all. And yet there is a secret freedom in the elegy, a freedom to wander back in time and reach forward, too, that tantalizes the mourner. To exercise this freedom, and yet to think that death may deprive us of it, is for Tennyson the most wrenching part of Hallam's tragedy. Unique to *In Memoriam* is the quality of its religious despair. As T.S. Eliot says in his great essay on the poem, 'to qualify' Tennyson's 'despair with the adjective "religious" is to elevate it above most of its derivatives. For *The City of Dreadful Night*, and *A Shropshire Lad*, and the poems of Thomas Hardy, are small in comparison with *In Memoriam*: It is greater than they and comprehends them' (1932b, 336).

No one would deny that the convergence of many voices in the 'con-

versation of mankind' is often more unsettling than consoling. Sometimes these voices remind us, in Rorty's phrase, that 'the things' we 'love
with all [our] heart' may not be 'central to the structure of the universe'
(1999, 20). Personality and justice cannot be preserved as ultimate
attributes of the Absolute, which in one philosopher's austere words 'is
not personal, nor ... moral, nor ... beautiful or true' (Bradley, 1893,
472). With equal force, however, a culture must also assert the counter-
truth that 'there is nothing sacred about universality' and that there 'is
no reason to be ashamed of, or downgrade, or try to slough off, your
Wordsworthian moments, your lover, your family, your pet, your favourite lines of verse, or your quaint religious faith' (Rorty, 1999, 13). We
yearn to 'see love, power and justice ... coming together deep down in
the nature of things' (Rorty, 1999, 19). But the tragedy of higher learning is its penchant for combining unwise power with powerless wisdom.
It is almost impossible to be personal and social, self-fulfilled and useful,
liberal and professional, at the same time. We can express this conclusion another way by saying that until philosophers become kings and
kings philosophers, it seems unlikely that the scholars and the lawyers,
the scientists and the politicians who dwell below Parnassus will ever
achieve the Muses' dream of holding the blissful and serviceable, the
pleasurable and the just, in a single vision.

Despite what Yeats says about 'monuments of unageing intellect,' few
professors can embalm and treasure up the precious life-blood of the
first book they write or the excitement of their first entrance into a new
and strange domain of scholarship and thought. By the time most scholars retire, their lifetime of investment in such specialties as the 'discarded image' of the Middle Ages or idealist philosophy among the
poets may show no more profit in the current market of ideas than their
bank accounts. But for the philosopher or literary critic who finds that
Plato or Wordsworth is no longer 'apparelled in celestial light,' often
the 'glory and the freshness of a dream' can still be recovered in a new
philosopher or poet, without devaluing the past. In a world where
'much abides,' though 'much is taken,' it is no dishonour to the venerable sages of the past that each generation of scholars should worship in
the holy city of Byzantium a fire-tongued band of new prophets and
seers.

The fragility of Socratic wisdom is an essential part of its charm. Built
in a region of earthquakes in Asia Minor, Byzantium is always in danger
of reverting to the intellectual barbarism of the sophists, who fight for
victory and profit rather than for truth. A society betrays art and culture

when it treats the scholar as a mere custodian or 'glorified clerk,' what one critic calls 'an obsessive archivist with a fetish for stasis' (O'Brien, 94). As prophet and scribe, the scholar must transmit as well as preserve wisdom. He must generate as well as control revolutionary energy in his discipline. A house of scholars is never a mere seminary or trade school, but a phantasmagoria of shifting towers and caves. Often the house disguises itself as a casino in which enterprising gamblers try to make their intellectual fortunes in a competitive market. In its mad dance of disciplines, the gambler's cave exhausts the freakishness of arbitrary caprice. Sometimes the murky labyrinth of the cave assumes the shape of an Ivory Tower, a home for reclusive scholars with the leisure to be wise. At other times the casino turns into an unsteady Tower of Babel, where energetic new thinkers compete to enter the conversation of mankind.

The end I fear most is the one foreseen by Matthew Arnold in his Preface to *Essays in Criticism*. He thinks the classical scholar he has unwittingly offended would be more tolerant of satiric high spirits 'if he considered that we are none of us likely to be lively much longer' (1961, 234). The prophecy comes to a focal point in the adjective 'lively.' The primary meaning is 'vivacious,' full of satiric energy and life. But Arnold once said his adversary's translation of Homer had 'no proper reason for existing.' Now he admits we all have a right to exist. The solemn justice of these thoughts brings to life the literal meaning of 'lively,' and precisely at that moment when it is in imminent danger of turning into its opposite, 'lethal' or 'deathlike.' The dark undertow comes to the surface in the next sentence, in which the prophet foresees that his 'vivacity is but the last sparkle of flame before we are all in the dark, the last glimpse of color before we all go into drab, – the drab of the earnest, prosaic, practical, literal future' (1961, 234). The variation on 'dark' and 'drab' sounds like a knell to toll Arnold back to everything he abhors in a world devoid of metaphor, poetry, and speculative thought. Instead of a free play of ideas, subtle and finely tuned, we hear only the monotonous thunder of philistines. The roaring of the young lions is magnificent but primitive. And they lack all sense of humour. In composing his own version of *The Dunciad*, Arnold's defence of learning aligns life, speculation, and wit against death, practicality, and boredom. In a dull world dominated by philistines 'we shall all yawn in one another's face,' Arnold gloomily foretells, 'with the dismallest, the most unimpeachable gravity' (1961, 234).

As lives wind down and friends begin to die, all pilgrims to Byzantium need to creep under some 'comfort' that will 'serve in a whirlwind'

(Hopkins, 'No worst, there is none,' l. 13). Though every parting is an image of death, George Steiner finds no consolation in the biblical paradox that we must die to live. Instead he brings the drama of his own retirement startlingly to life by putting unexpected pressure on the verb he uses to describe the breakup of a family: he says he is 'orphaned.'

> What I now experience of retirement from teaching has left me orphaned. My doctoral seminar in Geneva ran, more or less unbroken, for a quarter of a century. Those Thursday mornings were as near as an ordinary, secular spirit can come to Pentecost. By what oversight or vulgarization should I have been paid to become what I am? (19)

Instead of losing a young apprentice or offspring, the retiring professor loses an elder, a mentor or a parent. The horror of concluding is inseparable from any reflective teacher whose time to share ideas is limited and who receives as much from his doctoral students as they receive from him. Though Steiner is consoled by the thought that there is 'no craft more privileged' than 'the calling of the teacher,' he also knows that the satisfaction of being 'the servant, the courier of the essential,' can do 'nothing to alleviate death,' even though it may make a master spirit like Yeats or William James 'rage at its waste' (183–4).

Years ago I knew an aging scholar who waited with ever-decreasing hope for an honour and appointment that never came. He spent his last years chasing phantoms, fretting over every twitch of the leash, disappointed and embittered. I also know a more complacent, less ambitious scholar who, facing retirement, feared the loss of vocation as acutely as he felt the arctic wind catching his throat when he left Hart House and rounded the corner of Soldier's Tower. When Newman takes leave of Oxford in the *Apologia,* or when Paul Fussell describes the private ceremony of his bidding farewell to Rutgers, the experience is the same: each shows us his heart and pierces us.

> Before I left the New Brunswick–Princeton area for good, I felt the need for some sort of private ceremony recognizing the twenty-eight years I'd spent happily teaching at Rutgers. One Saturday, when I knew the classrooms would be unused and the campus largely deserted, I paid a sentimental visit to the places I'd taught in, pausing in each classroom to recall favorite students and moments of intellectual and psychological pleasure ... Every empty classroom echoes the substance of Ecclesiastes, warning the

happy, careless young that the days are coming when 'the keepers of the house shall tremble, and the strong men shall bow themselves, and the grinders cease because they are few, and those that look out of the windows be darkened.' (Fussell, 284–5)

A professor who meets his last graduate seminar or who lectures in a hall for the last time may reflect that the hour is at hand when he will do no more. As F.H. Bradley notices, 'after a certain age every milestone on our road is a grave-stone, and the rest of life seems a continuance of our own funeral procession' (1930, aphorism 70). But I am heartened by what Frank Kermode says. After giving up his chair at Cambridge, he reflects with Milton's Adam that 'some natural tears I dropped, but wiped them soon. The world was all before me where to choose' (258). Binding together luminous moments from the past are the intellectual pursuits and affections of a lifetime, which may attach themselves to a few well-loved authors and friends.

The motto of the university where I teach assumes that the realization of human potential is like the flourishing of Yeats's 'great-rooted blossomer': 'velut arbor aevo.' But uninhibited growth may be as undesirable in the human mind as in an unpruned vineyard. Apple and pear trees may have to be trimmed or transplanted as well as allowed to grow unrestrictedly as seedlings. A culture that idolizes growth by finding more value in 'half-acre tombs' than in well-wrought urns fails to see that arrival at a termination point is often arrival at the 'end' in another sense as well: it is the attainment of an objective or a goal. Unless our minds are bewitched by a commercial idol of unchecked growth in a free-market economy, we should be grateful that sometimes 'we can grow no further.' As Kadish perceives, 'we *have* grown, and to attain that state is at some stages the hoped-for end of a genuinely self-interested person.' Whenever a door shuts behind us, it is tempting to regret a road not taken or indulge a secret horror of concluding. 'We are idolators of the old,' as Emerson says. 'We cannot part with our friends. We cannot let our angels go' (1908c, 94). We are reluctant to admit that there are no birds this year in last year's nest. But instead of lingering in the ruins of old schools and shrines, I prefer to reflect with Kadish that 'in the very processes of higher education, and all through life, not only do possibilities close down; we are lost unless they do' (31).

Though all my mentors are dead now, I still reach out to them. Sometimes I take refuge in the thought that all things eventually 'merge into one, and a river runs through it. The river was cut by the world's great flood and runs over rocks from the basement of time. On some of the rocks are timeless raindrops. Under the rocks are the words, and some of the words are theirs' (Maclean, 113). But I can never decide whether the happy solemnity of the words is bracing or defeatist. Too often there rise unbidden in my mind, as in Yeats's elegy for Robert Gregory, the 'breathless faces' of two colleagues, Paul Gottschalk and John Finch, each a close friend and scholar who shared at an early age 'in the discourtesy of death.' Adding salt to the wound is the memory of a talented freshman I tactlessly offended two days before he died in the fire that killed John Finch. These events took place forty years ago. But they still take 'all my heart for speech.' Perhaps I see in every story of unfulfilled promise a figure of my own end. The longer I live the more afflicted I am with a panic-stricken sense of the transience and hence the seriousness of beautiful moments and people. The fair and shining promise of those I love and know best 'seriously, sadly runs away to fill the abyss's void with emptiness' (Robert Frost, 'West-Running Brook,' ll. 48–9).

The spectacle that haunts me is the gradual conquest or irremediable loss of that hunger for inquiry and shared response which make higher learning a climax and all enduring achievement in scholarship or science a harvest of passions sown in our youth. As I cast a last look back on this drama, I take comfort from the thought that Memory is the mother of the Muses because what we remember is saved from the universal cataract of change that spends to nothingness. Since our cultural memory is no longer part of the world that dies with each generation, it is also a release from 'the arctic winter' of Newman's passionless academy, 'an ice-bound, petrified, cast-iron University' (1856, 112) which sacrifices the personal influence of Socrates to sophistry and system. Hell is amnesia; the loss of connection; the nakedness and solitude of a discrete and therefore specious present. The Muses who inspire the arts and the scholars who study them are culture's way of making peace with time. Perhaps this is the meaning of Socrates' strange saying that all philosophy is a meditation on death. Northrop Frye has said that 'death is a leveler, not because everybody dies, but because nobody understands what death means' (1982, 320). But if our 'own resinous heart has fed' the Muses' sacred fires, then even the untouchable sadness of things can help us discover who we are.

Instead of asking the prophet's question, 'Why are you calling me, what would you have me do?,' the philosopher who becomes a corporation lawyer or the research scientist who becomes president of a drug company may have substituted for their daemon an idol of the marketplace or tribe. The scholar or scientist who rejects his calling may find much to deride in an apparent misfit like Henry Adams. As a Harvard professor, Adams has to peddle Clio, the Muse of history, to the university where he had studied as a youth without either improving his mind or enlarging his goals. The disenchanted Adams is not God's spy. But neither is he a master-leaver or deserter. As a scholar Adams is surely entitled to more hope and good cheer than he allows. If Caliban is bold enough to evolve from grub into butterfly, if he is moved by what Plato calls his master passion or 'mania,' he may be touched by the grace of 'sounds and sweet airs' (*The Tempest*, 3.3.145). When he explores and tries to lighten the burden of the mystery, clouds may open and riches drop upon him.

A scholar may be blamed for turning Parnassus into Babel. He may be accused of making a profession of his calling, of marketing the Muse or squandering the 'fountain light' of all his 'day.' But in defence of scholars it must be said that their lifelong devotion to study and teaching is often selfless and exacting. Our culture attaches little social prestige to learning, apart from the entry education provides into a few well-paying professions. And too often the perceived success or failure of education depends upon optional stop rules. Many students think their education a success because they stop learning at the point their financial future seems secure. They never push their learning far enough to realize that their education is a failure in the only sense that matters: they have wasted their gifts. To deny one's daemon is to quench the 'master light' of all one's 'seeing.' Like George Eliot's Lydgate, the medical pioneer in *Middlemarch* who dwindles into a fashionable London doctor, a scholar or scientist *manqué* has failed to do what he once meant to do. When old age wastes his generation, no monument will befriend him, as the urn befriends Keats.

Like Wordsworth's 'spots of time,' the 'call' to a life of scholarship or science may last no more than a moment, and so is the opposite of everlasting. But it is also in some sense eternal because the truths it invites us to create and the values it challenges us to renew are compressed into instants when our contact with the limits of what is conceivable rather than actual allows us to die on purpose to the rush of time. The best portrait of the scholar reflects a deep and backward-reaching intimacy. It mirrors the mentors we valued in our youth and now see clearly with the 'inward eye' of memory. If a scholar's ghost is to be saved from oblivion, it must salvage from its 'scrap of history' something less spectral than a photo album or an inscription on a gravestone, where 'only an attitude remains' (Larkin, 'An Arundel Tomb,' ll. 32–3, 42). 'What will survive of us' is not just 'love' but the sum total of transforming moments we ourselves have changed.

Notes

Chapter 1

1 In the power of 'personal knowledge' to reflect both the view-holder and his world-view, I try to give sharper definition to the term 'reflective inquiry.' Tom Pocklington and Alan Tupper use this second phrase to describe an alternative to 'frontier research' into ever smaller, esoteric topics in their book *No Place to Learn: Why Universities Aren't Working*. I borrow the first term, 'personal knowledge,' from Michael Polanyi's book on theories of knowledge in science: *Personal Knowledge: Towards a Post-Critical Philosophy*.

2 In his essay on 'Schopenhauer as Educator,' Nietzsche makes a radical distinction between the prophet and the scholar. Since Kant lacked Nietzsche's own prophetic genius, Nietzsche heretically denies that his scholarly precursor was a true philosopher. 'A scholar can never become a philosopher,' Nietzsche polemically asserts. 'Even Kant could not manage it and, despite the innate power of his genius, remained to the very end in chrysalis state. Those who think these words are unfair to Kant do not know that a philosopher is – not only a great thinker but a genuine human being' (quoted by Shattuck, 104).

3 Despite my sympathy with humanists who deplore the death of scholarship, I have no desire to write an elegy. Stanley Fish associates the lament for lost learning with a more inclusive genre which, like the Book of Jeremiah, tells 'a story of loss, the loss of a time when common values were acknowledged and affirmed by everyone' (250). His essay 'The Common Touch, or, One Size Fits All' places a long list of books on education in this category. As Northrop Frye has noted, myths of a deteriorating culture are as old as culture itself: 'Some time ago an archaeologist in the Near East dug up an inscription five thousand years old which told him that "children no longer obey their parents, and the end of the world is approaching"' (1963, 62).

Chapter 2

1 For a fascinating discussion of the complacent sciolists, who are often ignorant gamblers in disguise, readers may wish to consult Nassim Nicholas
Taleb's book, *Fooled by Randomness: The Hidden Role of Chance in the Markets and
in Life*. A mathematician and hedge fund operator, Taleb examines situations
in which randomness is mistaken for non-randomness: luck for skills, probability for certainty, conjecture for certitude, and so on.

Chapter 4

1 When choosing a line of inquiry, it is often difficult to decide whether a
scholar should imitate a monist like F.H. Bradley or a pluralist like Jeremy
Bentham or J.S. Mill. According to Isaiah Berlin, thinkers are either hedgehogs or foxes: they have either one big idea or many small ones. Like most
utilitarians, Bentham is a fox who breaks down complex problems into simple ones: in tackling each issue separately, he intentionally forgets the larger
picture. Bradley, by contrast, is a hedgehog whose habit of first asking and
answering the large questions may seem more worthy and ambitious. But by
attempting more than the fox, the hedgehog may achieve less in the end.

Only a hedgehog like the author of the 'Reality' section of Bradley's book
Appearance and Reality can seem wise. When the agile sceptic and fox of the
earlier 'Appearance' section of the book is being subtle and sly, he is not
being wise, and when he is wise he is not being subtle. The insights of a fox
are too multiple and shifty to seem wise. In Ecclesiastes, only the sheer persis-
tence of Koheleth's foxiness and disillusion can turn him into a wise prophet
of detachment without withdrawal.

Raising cosmic questions may humanize the scholar. But the big hedgehogs
– synthesizers like Bradley, Hegel, and company – may build a maze in which
even God would lose his way. Conversely, a scholar who is too myopic may lose
Blake's capacity 'to see a world in a grain of sand' and 'heaven in a wild
flower.' Whether as a fox or a hedgehog, scholars should try to unearth and
keep alive a consciousness of the unifying moments in their lives, and so
expand their experience of the mysteries that surround them.

2 As a highly self-critical authority, William James is a master of the telling
exception. According to James, the students who do best on the rapid reading
tests are not the scholars or scientists who produce the best intellectual work.
Most of the distinguished men in science and literature turn out be 'slow
readers' (1958, 97). Even the mystery of great teaching cannot be solved.
Sooner or later 'psychology and general pedagogy ... confess their failure, and

hand things over to the deeper springs of human personality to conduct the task' (1958, 81).

At the end of his chapter on 'Attention,' James concedes that the genius who approaches his subject obliquely, and whose 'best thoughts come through his mind-wanderings,' may be 'extremely efficient all the same' (1958, 86). Though the steady faculty of attention is a great boon, a genius can break all the rules and still succeed.

Scholastic aptitude and IQ tests seem to measure mainly the 'lightning automatic flashbacks' rewarded on TV shows and on the multiple-choice exams embraced most eagerly by professional schools. Instead of taking the scholar's 'brain temperature,' exams should also assess the 'slow and con-trolled response' that Frye associates with wisdom. 'If intelligence means insight,' then, as he explains, the high marks achieved on multiple-choice exams may be evidence of only a 'very limited intelligence ... Intensity of vision delays the automatic apparatus of response: the pure contemplative can't respond at all' (2000, 25).

Jacques Barzun's attack on multiple-choice testing is just as devastating as Frye's. Preferred by instructors too busy to evaluate thoughtful essays or inge nious solutions to problems in mathematics and science, these tests often lose in educational value what they gain in accuracy and ease of quick computer marking. According to Barzun, the worst feature of multiple-choice testing 'is that it breaks up the unity of what has been learned and isolates the pieces. In going through 50 or 100 questions nothing follows on anything else. It is the negation of the normal pattern-making of the mind. True testing issues a call for patterns, and this is the virtue of the essay examination.' Such an exam equips students to organize their ideas in ways that would help them diagnose an illness or train them to write a well-researched essay for the *New England Journal of Medicine.* By contrast, an excessive reliance on multiple-choice exams fragments the mind and turns the learners into 'cripples in consecu-tive thought' (Barzun, 1991, 35).

Equally critical of standardized tests is Mark Pattison, who claims that 'the beneficial stimulus which examination can give to study is an inverse ratio to the quality of intellectual exertion required' (1889, 1:491). Pattison deplores the use of tests as 'an instrument of mere torture which has made education impossible and crushed the very desire for learning' (1885, 303).

3 For a criticism of Oakeshott's conversational theory of learning quite differ-ent from my own, readers may wish to consult Geoffrey Hartman's book *A Critic's Journey.* In taking Richard Rorty to task for applying the metaphor of conversation to philosophy, Hartman fears that an informal style of philo-sophic writing may sacrifice intellectual rigour to accessibility to laymen

(1999, 14–16). But as Hartman himself prudently cautions, 'no order of discourse or institutional way of writing has a monopoly on either rigor or invention' (1999, 16).

4 Hazard Adams shrewdly reflects that many students who hope to make huge salaries as CEOs, lawyers, or physicians discover that these professions have been artificially limited in size to protect incomes and status. Such students may have to fall back on a broader foundation of learning to keep their minds taut and adaptable and to keep vocational training in its proper subordinate place. As a corrective to the mindless lemming-like rush of students into professions that do not really want them, universities should help learners adapt to a world of 'intellectual flexibility and new possibilities' (53).

5 An education in poetry and other arts may help slay some of the impure monsters spawned by alumni associations and academic administrators. Such monsters include athletic coaches who are paid more than professors, professional fund-raisers who will boost anything that calls itself academic, and the undignified practice of putting professors out for hire to big business. To keep 'the robe of the professor' as 'stainless as the ermine of the judge' (Flexner, 207), it may be necessary, Flexner thinks, to discourage professors of business, medicine, and law from taking jobs outside the university unless such moonlighting is required (in very modest measure) to advance professional research.

Chapter 6

1 Michael Oakeshott's rationalist bears the same relation to his true scholar that the sophists bear to Socrates. 'If by chance,' Oakeshott says, the *tabula rasa* of history 'has been defaced by the irrational scribblings of tradition-ridden ancestors, then the first task of the Rationalist must be to scrub it clean' (1991, 9). A sophist aspires to the efficiency of an engineer who masters techniques in order to satisfy a society's currently perceived needs. Anything that fails to achieve desirable short-term results he rejects as 'a piece of mysticism and nonsense' (1991, 9). Unfortunately, in turning politics and education into a useful science, the sophist may induce cultural amnesia, like Oakeshott's rationalist. Since sophists believe that all knowledge is reducible to a technique, they reject as spurious the practical knowledge that requires informed judgment, sympathy, or tact. Sophists seek to produce perfect clones of themselves rather than variations of a type. They teach the inert or dead knowledge of textbooks and cribs rather than the living knowledge of models supple enough to reinterpret the data and flexible enough to reinvent first principles and axioms.

2 Though Emerson identifies Oxford with the ethos of the English public schools, C.S. Lewis is presumably typical of many Oxford dons when he tries to dissociate the two and vows to make Magdalene College into something 'more than a country club for all the idlest "bloods" of Eton and Charter-house' (1966, 260). 'I really don't know what gifts the public schools bestow on their nurslings,' Lewis muses, 'beyond the mere surface of good manners: unless contempt for the things of the intellect, extravagance, insolence, self-sufficiency, and sexual perversion, are to be called gifts ... I sometimes wonder if this country will kill the public schools before they kill it.' Letter to his father, 3 November 1928 (1966, 261).

Chapter 7

1 Compared with Hegel, whose wrestling match with words is an epic struggle, J.S. Mill is a transparent, even arid thinker. Nevertheless, if we try to substitute a mere text-book summary of Mill's ideas for a treatise like his essay *On Liberty*, we lose half its power. To prove his ideas on our pulses, Mill animates his par-adoxes with vivid metaphors and locks his antitheses into memorable epi-grams. The culminating shock of the opening pages is that democracy's boast about 'self-government' should in fact be its logical opposite: a confession of enslavement. The 'self-government' praised in a democracy is 'not the gov-ernment of each by himself, but of each by all the rest' (1947, 4). Freedom switches without warning into a second tyranny, more subtle and insidious than the tyranny of kings.

A favourite axiom of conservative political theorists from Richard Hooker to Edmund Burke is that custom or art is man's second nature. Mill devises an aphorism of his own to feed off and subvert this axiom. A culture's so-called 'second nature' may appear in time so 'self-evident and self-justifying' that it is mistaken for the laws of nature itself. To combat the 'magical influence of cus-tom,' which bewitches the mind with 'an all but universal illusion' that the idols of its cave are reasonable and natural, Mill affirms that custom 'Is not only, as the proverb says, a second nature, but is continually mistaken for the first' (1947, 5). In denying that custom is a 'second nature,' Mill divests it of the glamour and prestige that belong more properly to the first or primal nature alone.

At the beginning of the second chapter, 'Of the Liberty of Thought and Discussion,' Mill asserts that no political state has a right to suppress the opin-ion of a single dissident. In one of his most memorable aphorisms Mill declares that 'if all mankind minus one, were of one opinion, and only one person were of the contrary opinion, mankind would be no more justified in silencing that one person, than he, if he had the power, would be justified in silencing mankind' (1947, 16). This great aphorism is less witty than judicious, for it has the majestic repose we associate with the Decalogue or the Golden Rule. The hinge of the sentence comes at the middle, at the pivotal point where a chiasmus ('all mankind minus one,' 'the single dissident,' 'the single man,' the 'rest of mankind') traces the sweep of a great parabola or reversing arc that allows Mill to include the mass of mankind and the dissenter in a common fate. One might expect that as a utilitarian Mill would be interested in calculating the quantity of support he can gather on behalf of any opinion. If happiness is the satisfaction of the greatest number of people, would the number of suppressed opinions not be worth determining? Should we not conduct a poll?

Mill anticipates and nullifies this democratic suggestion by turning the tables on the pollsters. If we are to resort to calculation, what has to be computed is not the loss to the dissenter himself but the loss to posterity. In a second more surprising aphorism Mill reminds us that 'the peculiar evil of silencing the expression of an opinion is, that it is robbing the human race; posterity as well as the existing generation' (1947, 16). The countless number of unborn thinkers who are denied the suppressed thought of a new Socrates or Jesus allows Mill to restore a quantitative measure of value. And yet if nothing less momentous than the well-being of posterity hangs in the balance, the near infinitude of persons who might be denied the chance of exchanging error for truth (or at least the chance of receiving a 'livelier impression of truth') also throws a monkey wrench into Mill's equation. For it might be argued that infinity minus one is equal to infinity itself. The suspicion that a statistical theory of value chills the heart and betrays the mind, however cherished it may be by Mill's fellow utilitarians, lingers on as the paragraph's disturbing afterthought. How could we dream that the return of Mill's mind upon itself could be compassed in a crib or textbook summary?

Chapter 8

1 Bill Readings cites the testimony of a Nobel laureate in physics, who says the purpose of undergraduate education is to introduce students to 'the culture of physics,' by which he means 'conversation among a community rather than ... a simple accumulation of facts.' According to its editor Diane Elam, Readings's own posthumously published book *The University in Ruins* is itself the result of conversations they had shared about new models of learning. 'Dwelling in the ruins of the University was not usually a silent occasion for Bill. Talk – whether it led to agreement or disagreement, whether it was serious or

silly – had everything to do with how he worked, thought, and envisioned a future for the University' (Readings, vii).

2 If two laws can be derived from each other, the logic of their deduction is circular, because it is difficult to determine which law is more fundamental. Though Newton derived his law of gravitation from Kepler's laws of planetary motion, Kepler's laws are usually derived today from Newton's law, since the latter seems more fundamental. But as one physicist says, 'Kepler's laws ... also govern the motion of electrons around the nucleus, where gravity is irrelevant. So there is a sense in which Kepler's laws have a generality that Newton's laws don't have' (Weinberg, 48). Because Newton's law can be derived from Kepler's and Kepler's from Newton's, the process of deduction is more like tracing a circle in two directions than drawing a vector with a one-directional arrow.

The axioms of a formal mathematical system must be taken on faith. Instead of trying to prove their axioms, Euclid and Newton use them as geographical markers. Or as David Berlinksi explains, axioms 'mark the spot where the system starts' (148). By contrast, the theorems that follow logically from the axioms, and which must be true if the axioms are true, 'mark the spot at which the system stops' (148). An axiom bears the same relation to a theorem as assumed statements bear to propositions that are proved. But 'proof' in a formal system is not a systematic testing of a proposition. It is rather 'the process of moving from one statement to another – inference on the wing' (148), so to speak.

Since no concept of meaning is allowed to contaminate either the axioms or the theorems, a mathematical system is as purely formal (or formally pure) as a Symbolist poem by Arthur Rimbaud or Paul Verlaine. Yet Berlinski admits that in urging upon his students a 'stern renunciation of meaning,' the mathematician Alonzo Church also 'demanded that [they] keep the intended interpretation of the symbols firmly in mind, and thus engaged [them] in a delicate form of double-think' (148).

A proof will seem merely trivial if the theorems a mathematician demonstrates have more obvious meanings than the axioms he takes on faith. Perhaps the meaningful mechanical statements Newton makes *about* a formal system like analytic geometry are more important than the theorems Descartes can prove *within* the formal system itself. Most physicists and intellectual historians would agree that Newtonian mechanics provides a meaningful world-view. It is also the world-view that Cartesian geometry seems invented to map out and explain. 'Within Newtonian mechanics,' as Berlinksi says, 'the parabola, the ellipse, and the hyperbola function as archetypes: They are the curves by which continuous change is classified' (262–3). Berlinski believes that the birth of modern science can be traced to the moment Newton gives mathematical voice to a physical concept. In the concrete idea of instantaneous velocity, the mathematical notion of the derivative of a function receives a local habitation and a name. 'When everything is brushed away, and the thing is reduced to its absolute essentials,' Berlinski concludes, 'it is the derivative that lies at the heart of the mechanical world view. Once it has been introduced all else follows' (262).

Chapter 9

1 In a lecture on 'Education and Power,' Claude Bissell usefully distinguishes between the cultures of persuasion and compulsion (1968, 244). And in *Philosophy and the Return to Self-Knowledge,* Donald Phillip Verene helpfully defines the culture of persuasion as a recovery of both rhetoric and poetry to restore knowledge of the self. For the logical coercion of a proof, a persuasive culture substitutes philosophical explanations. Encouraging a tolerance for paradox and two-way meanings, and for language that 'woos eloquence, not mathematics,' such a culture supports Richard Lanham's polemic against the 'gospel of normative clarity' (1974, 67, 137). In renewing respect for authority that is self-critical, it also challenges any assumption of dogmatic finality in higher learning.

2 Timothy Findley lists some essential attributes of a great mentor in his portrait of Thornton Wilder in *Inside Memory: Pages from a Writer's Workbook.* When commenting at length on one of Findley's manuscripts, Wilder seemed to the apprentice playwright 'all great teachers in one. The only assumptions he made were that knowledge was important and that I cared' (30). Like Socrates, Wilder recommends 'an impassioned questioning written from a position of bafflement – The character in a play must never know what is going to happen next' (30).

3 Years ago at an American university, a chairman whom I expected to be energetic and judicious turned out to be insecure and timid. Some administrators I feared would be merely abrasive proved to be models of tact and encouragement: they have steadily built department morale and proved highly effective in councils of power.

   Styles of leadership will vary, some favouring consensus and others more central control. To honour both teaching and research and strike a proper balance between the two, academic administrators should do everything in their power to keep alive and well the fragile but resilient network of rivalries over which they have the privilege to preside. Instead of using their power of office to pursue personal vendettas or grievances, they should help colleagues face the inevitable anxieties and risks of academic life with a measure of confidence and hope.

Though diplomacy is an art rather than a science, a few guidelines are useful. Scholars should be consulted in their areas of expertise and made to feel important. It is insulting for a department to vote on a candidate before a senior scholar in that area has time to submit written appraisals to the committee. Scholars should be encouraged to have long-term goals and should not be prodded into recycling ideas just for the sake of publishing on cue. All faculty members should also be satisfied that the summaries of their scholarship and teaching are fair and that the statistics are accurate. If some professors feel that evaluation forms make teaching a popularity contest and encourage inferior standards, they should be free to invite colleagues into their classrooms or even solicit the opinion of students they have previously taught. One guide to a professor's success as a graduate teacher is the quality of books his former students publish and the academic positions they eventually secure.

One can never overestimate the shortsightedness of registrars and deans who make the timetables too crowded and the schedules too cluttered for students to master a related group of subjects. O.B. Hardison speaks wittily of the 'curricular sedimentation' (37) that can be studied in the fossil record of any college calendar. Instead of rebuilding a curriculum from the foundation up, departments allow the sedimentation to accumulate from one generation to the next. More dangerously, in times of retrenchment and diminishing enrolments departments try to attract more students by annexing to their main circus some currently fashionable sideshow. In John Dewey's words, 'it is our American habit if we find the foundations of our educational structure unsatisfactory to add another story or wing. We find it easier to add a new study or course or kind of school than to reorganize existing conditions so as to meet' changing needs (quoted by Charlton, 2).

4 Nowhere are the hazards of higher learning greater than in the precarious rituals of scholarly succession. The tattered remnants of a genuine apprenticeship system can be dimly discerned in the faculty seminar, the teaching workshop, or the Society of Fellows at Harvard, which Lawrence Lowell founded as an alternative to the PhD method of certifying successors (Barzun, 1968, 255). But since many of these remnants of a self-renewing apprenticeship system have often fallen into disuse, the awarding of tenure in a university tends to be the riskiest of all enterprises. As Barzun says, 'anxiety to find and fear of making a mistake have surrounded professor-hunting with all the apprehensions of indissoluble marriage; for tenure is marriage, and without any pledge to obey and cherish' (1968, 256). Far from maximizing the likeli-

hood of a university's reproducing itself responsibly, the pathological 'fussiness of choice' (Barzun, 1968, 256) in tenure decisions actually increases the risk of making serious mistakes.

Whereas critics of tenure may see it as a 'socialist conspiracy to protect incompetence' (Damrosch, 53), its defenders view it as a marriage contract or as an invitation to join an extended family. Unfortunately, very few academic departments are happy families. Because information is leaked from the most confidential meetings and appointment committees are free to ignore the advice of any expert in the field, few veterans would dispute Damrosch's cynical claim that 'clubhouse mechanisms' continue to operate: 'what we have in academia is not democracy as we know it today, but a plutocracy with a sugar coating of Stalinism' (70).

The politics of the department is not the politics of the household, which is the analogy Aristotle chooses for his city state. Nor is it the politics of the private social club or law firm. Though it is fashionable to call a faculty a 'team,' few 'scholars actually work together even in their teaching, still less in their research' (Damrosch, 55).

Both Northrop Frye and George Santayana compare the scholar to a polyp, a solitary worker unaware of the coral island he is building (Damrosch, 34-5). Such polyps have obvious affinities with graduate students, who often inhabit a desert world of monks or nomads whose 'ascetic practices of self-discipline and meditative solitude' seem light-years removed from the vortex or whirl of their 'jet-setting' supervisors (Damrosch, 19). Damrosch thinks the university paradoxically unites extremes of corporate identity and isolation. It incongruously offers the continuity of a 'lifetime association' to scholarly 'pieceworkers' or free agents, whose academic offices and studies are a 'cottage industry in itself' (56–7).

Chapter 10

1 Whether an aphorism is wise, witty, or profound, it belongs to a distinct genre of discourse. But except for Francis Bacon's distinction between aphorisms that are 'peremptory' (or 'magistral') and probative, there is no taxonomy of aphorisms.

Characteristic of the wise or judicious aphorism is the axis of truth which balances a reader between poles of judgment and sympathy, satire and compassion. Some scholars are surprised that George Eliot should model her unflattering portrait of the scholar Casaubon on her friend Mark Pattison, who had married a woman twenty-seven years his junior, as had Edward Casaubon, whose Renaissance namesake, Isaac, was the subject of a biography by Pattison. What such critics fail to see, however, is that no sooner has Eliot begun to skewer Casaubon on the spit of her exquisite and refined satiric intelligence than she stands back to view him with compassion and understanding. 'For my part,' she says, 'I am very sorry for him' (206). In a form of 'satire *manqué*' or satire foiled, the wise aphorist reaches down to rescue Casaubon from her own incisive blows.

Whereas a witty aphorism explains the marvellous in terms of the commonplace, a profound aphorism explains the commonplace in terms of the marvellous, even at the risk of retreating into a mist that God himself would think twice before penetrating. In 'The Latest Decalogue,' Arthur Hugh Clough replaces the Mosaic law with the aphorisms of a self-serving pragmatist: 'No graven images may be / Worshipped, except the currency' (ll. 3-4). In his essay 'Compensation,' Emerson substitutes for the moral law the aphorisms of a profound but disturbing cosmic law, according to which 'the dice of God are always loaded' and 'there is a crack in every thing God has made' (1908c, 76, 80).

Since Emily Dickinson believes 'the unknown is the largest need of intellect' (quoted by Kazin, 155), she is a reluctant aphorist, afraid to formulate axioms that presume to know God as well as God knows himself. Like Christina Rossetti, another poet of grievous reticences, Dickinson believes there is a way things are, but that only fools presume it is ultimately knowable or known. 'She would have liked Simone Weil's "Attentiveness without an object is the supreme form of prayer"' (Kazin, 150).

Browning is a silent aphorist, for whom the Tao that can be spoken is not the real Tao. The aphorism in the coda of 'Love among the Ruins,' the resounding affirmation that 'Love is best' (l. 84), is not the oracle the lover has hoped to utter, nor is it the discovery the reader has been waiting to hear. Having substituted an aphorism for an insight, Browning has to intimate the lover's deepest meaning in another lyric, 'Two in the Campagna.'

2 When Oscar Wilde observes that 'to Parnassus there is no primer and nothing that one can learn is ever worth learning' (1968, 3), he may mean that knowledge of Plato or Keats can be acquired only through a temperamental affinity for philosophy or poetry. Or he may mean that there are no rules for having insights into Plato and Keats. All that can be taught is a method or a craft, like the Socratic method of inquiry and response in which the true teacher is the learner. But even if the most important knowledge is innate rather than acquired, education (as its etymology suggests) can at least 'draw out' and refine such knowledge. If all learning is self-learning, then its mysteries are profaned whenever teaching becomes 'the inculcation of the incomprehensible into the indifferent by the incompetent,' as John Maynard Keynes pithily puts it (quoted by Charlton, 26).

Plato says of knowledge what Keats says of poetry: it is a recollection, and should strike the mind as 'almost a remembrance.' Even after the propositions in Spinoza's *Ethics* and Euclid's geometry have faded from memory, I still retain a clear impression of the pleasure I once took in each concise and elegant demonstration. The half-intuitive, half-conscious ability to think, to feel, and to recognize in subjects that were once studied the aesthetic equivalent of intellectual cogency remains clear and distinct when all else fades. Such stubborn survivals are what Oakeshott means, I think, by the incommunicable 'shadow of lost knowledge' (1989, 61).

3 The values that appeal to Frost find expression in a language of two-way meaning and paradox. Such duality or openness is what Geoffrey Hartman has called 'a counter-propaganda,' disclosing 'that the whole is vaguer than its parts' and that the pretence of unity is only 'magic glue or myth' (1999, xxviii). The culture that both technological society and the modern university have betrayed is a culture of what Kant would call purposeless design. Though this culture has 'its own form of antimonumentalism' (Hartman, 1999, xxviii), its aesthetic of play and implicit reasoning has been largely displaced by a culture of explicit reasoning, logical coercion, and proof.

The discipline imposed by the frugal tetrameter quatrains of *In Memoriam* supports Hartman's claim that literary language is 'the most ascetic, critical, and commonplace basis of art' (1999, xxviii). When we recall the double meaning of 'bond' (both a fetter that constrains and a tie that binds freely), we can better understand why intellectually demanding disciplines are often the opposite of coercive and why any defence of humane learning must assume the form of a secular theology whose service to truth is performed with perfect freedom because it is also an act of love.

Chapter 11

1 A safe method of subverting authority to renew tradition is to challenge a community's axioms of faith within situations that make the challenge culturally innocuous. An example Hillis Miller considers is the English social novel, which has traditionally maintained the authority of a culture's shared norms and values by creating 'groups of people who live by the same fictions' (105). But social novels also delight in the playful undoing or unravelling of such fictions. They often expose the fictive status of a character's identity even as they posit that identity as a first axiom of faith. In the last paragraph of *David Copperfield*, for example, Dickens's characters are at once melting shadows – such stuff as dreams are made on – and the enduring substance of the narrator's vision: 'O Agnes, O my soul, so may thy face be by me when I close my life

indeed; so may I, when realities are melting from me, like the shadows which I now dismiss, still find thee near me, pointing upward!'

A novel like *Vanity Fair* or a short story like Alice Munro's 'The Progress of Love' may also demolish the illusion of a solid and abiding self at the centre of each important character. As Thackeray lets us see behind the masks of all his off-stage actors and actresses, he allows us a glimpse of their bad faith and casuistry. And in 'The Progress of Love' Alice Munro subversively intimates that the legends Euphemia and her mother perpetuate about parents they irrationally love or hate are, in truth, only favoured fictions – or lies to live by.

Miller's main point, however, is that in telling such secrets Thackeray and Munro are merely 'destroying' what is 'already destroyed, in order to preserve the illusion that it is still intact.' As he understands, 'all men and women living within a culture accepting a certain notion of character have an uneasy feeling that their belief in character, even their belief in their own character, may be confidence in an illusion.' The purpose of subverting a fiction no longer believed in is 'to assuage this covert suspicion by expressing it overtly, in a safe region of fiction' (106). Only after its authority has been challenged and dissolved can the novel's convention of an abiding self be reexamined and perhaps restored.

2 Harold Bloom amplifies his meaning in a response to the skeptic who questions such love. 'Well, what is teacher's love? In the English-speaking academic world, closely ruled by campus Puritans, we now have knitting-circles of Madame Defarges, sadistically awaiting the spectacle of the guillotine, fit punishment for 'sexual harassment,' that poor parody of the Socratic Eros.' Though the feminist scholar Naomi Wolf has recently aspired to the role of Madame Defarges, Bloom believes that 'an eros more dualistic even than that of Socrates is appropriate, indeed essential, for effective teaching' (2002, 24).

The last word on this tricky subject belongs to George Steiner, who admits that the dangers and privileges of teaching are boundless. Since 'the pulse of teaching is persuasion,' it often verges on and invites an erotic response. 'Eroticism,' says Steiner, 'covert or declared, fantasized or enacted, is inwoven in teaching, in the phenomenology of mastery and discipleship. This elemental fact has been trivialized by a fixation on sexual harassment. But it remains central. How could it be otherwise?' (26).

The abuse of eros is no argument against its right use. In an age of grim evangelicals, Matthew Arnold adds his weight to the lighter scale, balancing out the forces of sweetness and light, those gracious fugitives from the prison of Puritanism. The great enemy of eros is the sophist, responsible for what Steiner calls 'the mercantilization of the Master's calling' (20), and ever suspicious of any power or persuasive agency not reducible to rules.

Works Cited

Abrams, M.II. 1971. *Natural Supernaturalism: Tradition and Revolution in Romantic Literature.* New York: W.W. Norton.

Adams, Hazard. 1988. *The Academic Tribe.* 2nd ed. Urbana, Chicago: University of Illinois Press.

Adams, Henry. 1918. *The Education of Henry Adams.* New York: Modern Library.

Agee, James. 1939. *Three Tenant Farmers: Let Us Now Praise Famous Men.* Boston: Houghton Mifflin Co.

Annan, Noel. 1971. *What Is a University For Anyway?* Toronto: University of Toronto Press.

– 1999. *The Dons: Mentors, Eccentrics and Geniuses.* Hammersmith, London: HarperCollins.

Arnold, Matthew. 1932. *Culture and Anarchy.* Ed. J. Dover Wilson. Cambridge: Cambridge University Press.

– 1961. *Poetry and Criticism of Matthew Arnold.* Ed. A. Dwight Culler. Boston: Houghton Mifflin.

Bacon, Francis. 1955. *Selected Writings of Francis Bacon.* Ed. Hugh G. Dick. New York: Random House.

Barzun, Jacques. 1964. *Science: The Glorious Entertainment.* New York: Harper and Row.

– 1968. *The American University: How It Runs, Where It Is Going.* New York, Evanston, London: Harper and Row.

– 1991. *Begin Here: The Forgotten Conditions of Teaching and Learning.* Chicago, London: University of Chicago Press.

Bate, W. Jackson. 1975. *Samuel Johnson.* New York: Harcourt Brace Jovanovich.

Bate, W. Jackson, Michael Shinagel, and James Engell, eds. 1991. *Harvard Scholars in English, 1890 to 1990.* Cambridge, Mass.: Harvard University Press.

Berlin, Isaiah. 1998. *The Proper Study of Mankind: An Anthology of Essays.* Ed. Henry Hardy and Roger Hausheer. New York: Farrar, Straus and Giroux.

Berlinski, David. 1988. *Black Mischief: Language, Life, Logic, Luck.* New York: Harcourt Brace.

Bissell, Claude. 1968. *The Strength of the University.* Toronto: University of Toronto Press.

– 1974. *Halfway Up Parnassus: A Personal Account of the University of Toronto 1932–1971.* Toronto: University of Toronto Press.

Bloom, Allan. 1987. *The Closing of the American Mind: How Higher Education Has Failed Democracy and Impoverished the Souls of Today's Students.* New York: Simon and Schuster.

– 1993. *Love and Friendship.* New York: Simon and Schuster.

Bloom, Harold. 1998. *Shakespeare: The Invention of the Human.* New York: Riverhead, 1998.

– 2002. *Genius: A Mosaic of One Hundred Exemplary Creative Minds.* New York: Warner Books.

Bok, Derek. 1986. *Higher Learning.* Cambridge, Mass.: Harvard University Press.

– 1993. *The High Cost of Talent: How Executives and Professionals Are Paid and How It Affects America.* New York: Free Press.

Boone, Joseph Allen. 1987. *Tradition Counter-Tradition: Love and the Form of Fiction.* Chicago and London: University of Chicago Press.

Booth, Stephen. 1998. *Precious Nonsense: The Gettysburg Address, Ben Jonson's Epitaphs on His Children, and 'Twelfth Night.'* Berkeley, Los Angeles: University of California Press.

Bradley, F.H. 1876. *Ethical Studies.* London and Oxford: Oxford University Press.

– 1883. *The Principles of Logic.* London: Kegan Paul, Trench.

– 1893. *Appearance and Reality: A Metaphysical Essay.* Oxford: Clarendon Press.

– 1930. *Aphorisms.* Oxford: Clarendon Press.

Bromwich, David. 1992. *Politics by Other Means: Higher Education and Group Thinking.* New Haven: Yale University Press.

Brontë, Emily. 1963. *Wuthering Heights.* Ed. William M. Sale, Jr. New York: Norton.

Browning, Robert. 1981. *Robert Browning: The Poems.* Ed. Thomas J. Collins and John Pettigrew. 2 vols. New Haven and London: Yale University Press.

Burton, Richard F. 1964. *Personal Narrative of a Pilgrimage to Al-Madinah and Meccah.* 2 vols. New York: Dover.

Byatt, A.S. 1992. 'The Conjugial Angel.' *Angels and Insects: Two Novellas.* New York: Random House.

Callaghan, Barry. 1998. *Barrelhouse Kings.* Toronto: McArthur and Co.

Cameron, J.M. 1978. *On the Idea of a University*. Toronto: University of Toronto Press.

Carlyle, Thomas. 1970. *Sartor Resartus and Selected Prose*. New York: Holt, Rinehart and Winston.

Cassirer, Ernst. 1961. *The Logic of the Humanities*. Trans. Clarence Smith Howe. New Haven: Yale University Press.

Chadwick, Owen. 1983. *Newman*. Oxford: Oxford University Press.

Charlton, James, ed. 1994. *A Little Learning Is a Dangerous Thing: A Treasury of Wise and Witty Observations for Students, Teachers, and Other Survivors of Higher Education*. New York: St Martin's Press.

Chomsky, Noam. 1991. *Necessary Illusions: Thought Control in Democratic Societies*. Toronto: House of Anansi Press.

Coleridge, Samuel Taylor. 1905. *Aids to Reflection*. Edinburgh: John Grant.

Collingwood, R.G. 1939. *Autobiography*. London: Oxford University Press.

Comenius, J.A. 1968. *John Amos Comenius on Education*. Intro. Jean Piaget. New York: Teachers College Press of Columbia University.

Culler, A. Dwight. 1955. *The Imperial Intellect: A Study of Newman's Educational Ideal*. New Haven: Yale University Press.

Damrosch, David. 1995. *We Scholars: Changing the Culture of the University*. Cambridge, Mass.: Harvard University Press.

Denby, David. 1996. *Great Books*. New York: Simon and Schuster.

De Tocqueville, Alexis. 1998. *Democracy in America*. Trans. Henry Reeve. Crib State, Ware, Hertfordshire: Wordsworth Editions.

Dickinson, Emily. 1951. *The Complete Poems of Emily Dickinson*. Ed. Thomas H. Johnson. Boston and Toronto: Little, Brown and Co.

Dixon, William Macneile. 1937. *The Human Situation*. London: Penguin.

Donoghue, Denis. 1992. 'Mister Myth.' *New York Review of Books*. 9 April.

– 1997. 'The Practice of Reading.' *What's Happened to the Humanities?* Ed. Alvin Kernan. Princeton: Princeton University Press, 122–40.

Dworkin, Ronald. 2001. *Sovereign Virtue: The Theory and Practice of Equality*. Cambridge, Mass.: Harvard University Press.

Eliot, George. 1956. *Middlemarch*. Ed. Gordon S. Haight. Boston: Houghton Mifflin.

Eliot, T.S. 1932a. 'Francis Herbert Bradley.' *Selected Essays*. London: Faber and Faber, 444–55.

– 1932b. '*In Memoriam.*' *Selected Essays*, 328–38.

– 1936. *T.S. Eliot: Collected Poems 1909–1935*. New York: Harcourt, Brace.

– 1942. *The Classics and the Man of Letters*. London, New York, Toronto: Oxford University Press.

– 1943. *Four Quartets*. New York: Harcourt, Brace.

– 1948. *Notes towards the Definition of Culture*. London: Faber and Faber.

Emerson, R.W. 1908a. 'Universities.' *English Traits, Representative Men and Other Essays by Ralph Waldo Emerson*. London: J.M. Dent, 99–106.

– 1908b. 'Plato: New Readings.' *English Traits*, 193–8.

– 1908c. 'Compensation.' *Emerson's Essays*. First series. New York: Thomas Y. Crowell, 69–95.

– 1908d. 'Friendship.' *Emerson's Essays*, 142–62.

– 1908e. 'Circles.' *Emerson's Essays*, 220–36.

Engell, James, and David Perkins. 1988. *Teaching Literature: What Is Needed Now*. Harvard English Studies 15. Cambridge, Mass.: Harvard University Press.

Findley, Timothy. 1990. *Inside Memory: Pages from a Writer's Workbook*. Toronto: HarperCollins.

Finley, M.I. 1964. 'Crisis in the Classics.' *Crisis in the Humanities. The Politics of Liberal Education*. Ed. J.H. Plumb. Harmondsworth: Penguin.

Fish, Stanley. 1992. 'The Common Touch, or, One Size Fits All.' *The Politics of Liberal Education*. Ed. Darryl J. Gless and Barbara Herrnstein Smith. Durham, NC, and London: Duke University Press, 241–66.

Fleishman, Avrom. 1998. *The Condition of English: Literary Studies in a Changing Culture*. Westport, Conn.: Greenwood Press.

Flexner, Abraham. 1930. *Universities, American, English, German*. New York: Oxford University Press.

Forster, E.M. 1960. *The Longest Journey*. Harmondsworth: Penguin.

Foucault, Michel. 1973. *The Order of Things: An Archaeology of the Human Sciences*. New York: Vintage.

Frost, Robert. 1961. *The Poetry of Robert Frost*. Ed. Edward Connery Lathem. New York: Holt, Rinehart and Winston.

– 1968. *Selected Prose of Robert Frost*. Ed. Hyde Cox and Edward Connery Lathem. New York: Collier.

– 1981. *Robert Frost and Sidney Cox: Forty Years of Friendship*. Ed. William R. Evans. Hanover, NH: University Press of New England.

Frye, Northrop. 1963. *The Educated Imagination*. Toronto: CBC Publications.

– 1964. 'The Problem of Spiritual Authority in the Nineteenth Century.' *Essays in English Literature from the Renaissance to the Victorian Age, Presented to A.S.P. Woodhouse*. Ed. Millar MacLure and F.W. Watt. Toronto: University of Toronto Press, 304–19.

– 1967. 'The Instruments of Mental Production.' *The Knowledge Most Worth Having*. Ed. Wayne C. Booth. Chicago: University of Chicago Press, 59–83.

– 1982. *The Great Code: The Bible and Literature*. Toronto: Academic Press Canada.

– 1988. *On Education*. Markham, Ont.: Fitzhenry and Whiteside.

– 1990. *Words with Power.* New York: Harcourt Brace.
– 1991. *The Double Vision: Language and Meaning in Religion.* Toronto: University of Toronto Press.
– 1991–2. *Northrop Frye Newsletter.* Vol. 4, no. 1. Winter.
– 2000. *Northrop Frye Newsletter.* Vol. 8, no. 2. Fall.
Fussell, Paul. 1996. *Doing Battle: The Making of a Skeptic.* Boston: Little, Brown.
Garber, Marjorie. 2001. *Academic Instincts.* Princeton: Princeton University Press.
Getman, Julius. 1992. *In the Company of Scholars: The Struggle for the Soul of Higher Education.* Austin: University of Texas Press.
Giamatti, A. Bartlett. 1981. *The University and the Public Interest.* New York: Atheneum.
Gless, Darryl J., and Barbara Herrnstein Smith. 1992. *The Politics of Liberal Education.* Durham, NC: Duke University Press.
Gooch, Paul W. 1996. *Reflections on Jesus and Socrates: Word and Silence.* New Haven: Yale University Press.
Gould, Stephen Jay. 2003. *The Hedgehog, the Fox, and the Magister's Pox: Mending the Gap between Science and the Humanities.* New York: Harmony Books.
Graff, Gerald. 1987. *Professing Literature: An Institutional History.* Chicago: University of Chicago Press.
Grafton, Anthony. 2001. 'A Passion for the Past.' *New York Review of Books.* Vol. 48, no. 4, 47–54.
Grant, George. 1980. 'The Battle between Teaching and Research.' *Globe and Mail.* 28 April 7.
– 1998. *The George Grant Reader.* Ed. William Christian and Sheila Grant. Toronto: University of Toronto Press.
Greenblatt, Stephen. 1991. *Marvelous Possessions: The Wonder of the New World.* Chicago: University of Chicago Press.
Grote, George. 1846–56. *A History of Greece, A New Edition.* 12 vols. London: Murray.
Grote, John. 1865. *Exploratio Philosophica.* 2 vols. Cambridge: Cambridge University Press.
Hampshire, Stuart. 1967. 'Commitment and Imagination.' *The Morality of Scholarship.* Ed. Max Black. Ithaca: Cornell University Press, 31–55.
Hardison, O.B., Jr. 1972. *Toward Freedom and Dignity: The Humanities and the Idea of Humanity.* Baltimore: Johns Hopkins University Press.
Harrison, Frederic. 1911. *Autobiographic Memoir.* London: Oxford University Press. 2 vols.
Hart, Jeffrey. 2001. *Smiling through the Cultural Catastrophe: Toward the Revival of Higher Education.* New Haven and London: Yale University Press.
Hartman, Geoffrey. 1980. *Criticism in the Wilderness: The Study of Literature Today.* New Haven: Yale University Press.

– 1999. *A Critic's Journey: Literary Reflections 1958–1998.* New Haven: Yale University Press.

Heaney, Seamus. 2002. *Finders Keepers: Selected Prose 1971–2001.* London: Faber and Faber.

Hegel, Georg W.F. 1967. *The Phenomenology of Mind.* Trans. J.B. Baillie. New York: Harper Torchbooks.

Heidegger, Martin. 1968. *What Is Called Thinking?* Trans. J. Glenn Gray. New York: Harper and Row.

Highet, Gilbert. 1954. *Man's Unconquerable Mind.* New York: Columbia University Press.

Hollander, John. 1975. *Vision and Resonance: Two Senses of Poetic Form.* New York: Oxford University Press.

Honan, Park. 1999. *Shakespeare: A Life.* Oxford: Oxford University Press.

Hopkins, G.M. 1967. *Poems.* Ed. W.H. Gardner and H.N. Mackenzie. 4th ed. London: Oxford University Press.

Hughes, Thomas. 1994. *Tom Brown's Schooldays.* Harmondsworth: Penguin.

Huxley, T.H. 1904–25. *Collected Essays.* 9 vols. London: Macmillan.

Ignatieff, Michael. 1998. *A Life of Isaiah Berlin.* Harmondsworth: Penguin.

Innis, Harold. 1951. *The Bias of Communication.* Toronto: University of Toronto Press.

Irving, Allan. 2000. 'When Less Is More: In an Age of Academic "Stars," a "Good Enough" Professor Seeks Solace in the Words of Beckett and the Ambiguities of Art.' *University of Toronto Bulletin.* Nov. 7. 13 November, 16.

Jackson, H.J. 2001. *Marginalia: Readers Writing in Books.* New Haven and London: Yale University Press.

Jackson, Kevin. 1999. *Literary Forms: A Guide to Literary Curiosities.* London: Macmillan.

James, Henry. 1962. 'Daisy Miller: A Study.' *Henry James: 'The Turn of the Screw' and Other Short Stories.* New York: Signet.

James, William. 1958. *Talks to Teachers on Psychology; and to Students on Some of Life's Ideals.* New York: W.W. Norton.

– 1961. *The Selected Letters of William James.* New York: Anchor.

Jarrell, Randall. 1954. *Pictures from an Institution: A Comedy.* New York: Alfred A. Knopf.

Johnson, Fenton. 1996. *Geography of the Heart: A Memoir.* New York: Washington Square Press.

Johnson, Samuel. 1958. *Samuel Johnson. Rasselas, Poems, and Selected Prose.* Ed. Bertrand H. Bronson. New York and Toronto: Rinehart.

Jowett, Benjamin, trans. and ed. 1871. *The Dialogues of Plato.* 4 vols. Oxford: Oxford University Press.

Joyce, James. 1958. *Dubliners.* New York: Viking.

Kadish, Mortimer R. 1991. *Toward an Ethic of Higher Education.* Stanford: Stanford University Press.

Kazin, Alfred. 1997. *God and the American Writer.* New York: Alfred A. Knopf.

Keats, John. 1959. *Selected Poems and Letters of John Keats.* Ed. Douglas Bush. Boston: Houghton Mifflin.

Keble, John. 1838. 'Review of Lockhart's *Life of Sir Walter Scott.*' *British Critic and Quarterly Theological Review.* Vol. 24.

Kermode, Frank. 1995. *Not Entitled: A Memoir.* New York: Farrar, Straus and Giroux.

Kernan, Alvin, ed. 1997. *What's Happened to the Humanities?* Princeton: Princeton University Press.

– 1999. *In Plato's Cave.* New Haven: Yale University Press.

Kierkegaard, Søren. 1965. *The Concept of Irony.* Trans. Lee M. Capel. Bloomington: Indiana University Press.

Kristeva, Julia. 1989. *Black Sun: Depression and Melancholia.* Trans. L.S. Roudiez. New York: Columbia University Press.

Kuhn, Thomas S. 1962. *The Structure of Scientific Revolutions.* Chicago: University of Chicago Press.

Lange, John. 1970. *The Linguistic Paradox: An Inquiry Concerning the Claims of Philosophy.* Princeton: Princeton University Press.

Langer, Susanne K. 1942. *Philosophy in a New Key: A Study in the Symbolism of Reason, Rite, and Art.* Cambridge: Harvard University Press.

Lanham, Richard A. 1974. *Style: An Anti-Textbook.* New Haven: Yale University Press.

– 1983. *Literacy and the Survival of Humanism.* New Haven: Yale University Press.

– 1992. 'The Extraordinary Convergence: Democracy, Technology, Theory, and the University Curriculum.' *The Politics of Liberal Education.* Ed. Darryl J. Gless and Barbara Herrnstein Smith. Durham, NC, and London: Duke University Press, 33–56.

Larkin, Philip. 1988. *Philip Larkin: Collected Poems.* Ed. Anthony Thwaite. London: Farrar, Straus, Giroux.

Leavis, F.R. 1943. *Education and the University: A Sketch for an 'English' School.* London: Chatto and Windus.

– 1998. *The Critic as Anti-Philosopher.* Ed. G. Singh. Chicago: Ivan R. Dee.

Lewis, C.S. 1966. *Letters of C.S. Lewis.* Ed. W.H. Lewis. Orlando, Florida: Harcourt.

– 1967. *The Discarded Image: An Introduction to Medieval and Renaissance Literature.* Cambridge: Cambridge University Press.

Maclean, Norman. 1992. *A River Runs through It and Other Stories*. New York: Pocket Books.

Magee, Bryan. 1997. *Confessions of a Philosopher: A Journey through Western Philosophy*. New York: Random House.

Mandelbaum, Maurice. 1971. *History, Man, and Reason: A Study in Nineteenth-Century Thought*. Baltimore: Johns Hopkins University Press.

Marchand, Philip. 1989. *Marshall McLuhan: The Medium and the Messenger*. New York: Ticknor and Fields.

Maritain, Jacques. 1963. 'The Seven Misconceptions of Modern Education.' *The Teacher and the Taught*. Ed. Ronald Gross. New York: Dell, 157–77.

Menand, Louis. 2001a. 'The Socrates of Cambridge.' *New York Review of Books*. Vol. 48. 26 April, 52–5.

– 2001b. 'College: The End of the Golden Age.' *New York Review of Books*. Vol. 48. 18 October 44–7.

Merton, Thomas. 1997. *Entering the Silence*. Vol. 2 (1941–52) of *The Journals of Thomas Merton*. Ed. Jonathan Montaldo. San Francisco: HarperCollins.

Metcalf, John. 2003. *An Aesthetic Underground: A Literary Memoir*. Toronto: Thomas Allen.

Meyers, Jeffrey. 1996. *Robert Frost: A Biography*. New York: Houghton Mifflin.

Mill, John Stuart. *On Liberty*. 1947. Ed. Alburey Castell. New York: Appleton-Century-Crofts.

– 1950. *Mill on Bentham and Coleridge*. Ed. and intro. F.R. Leavis. London: Chatto and Windus.

– 1974. *System of Logic: Ratiocinative and Inductive*. Vol. 8 of *Collected Works of John Stuart Mill*. Ed. J.M. Robson and R.F. McRae. Toronto: University of Toronto Press.

– 1981. *Autobiography and Literary Essays*. Vol. 1 of *Collected Works of John Stuart Mill*. Ed. J.M. Robson and Jack Stillinger. Toronto: University of Toronto Press.

Miller, Hillis. 1988. 'The Function of Rhetorical Study at the Present Time.' *Teaching Literature: What Is Needed Now*. Cambridge, Mass.: Harvard University Press, 87–109.

Murdoch, Iris. 1992. *Metaphysics as a Guide to Morals*. New York: Allen Lane.

Nagel, Thomas. 1986. *The View from Nowhere*. New York: Oxford University Press.

Nasar, Sylvia. 1998. *A Beautiful Mind: The Life of Mathematical Genius and Nobel Laureate John Nash*. New York: Simon and Schuster.

Newman, John Henry. 1856. *The Office and Work of Universities*. London: Longman, Brown, Green, and Longmans.

– 1887. *Sermons Preached before the University of Oxford between 1826 and 1843*. London: Rivingtons.

– 1959. *The Idea of a University*. Garden City, NJ: Image Books.

– 1968. *Apologia Pro Vita Sua*. Ed. D.J. DeLaura. New York: W.W. Norton.

Nietzsche, Friedrich. 1966. *Beyond Good and Evil: Prelude to a Philosophy of the Future*. Trans. Walter Kaufmann. New York: Vintage.

Nock, Albert Jay. 1964. *The Memoirs of a Superfluous Man*. Chicago: Henry Regnery Co.

Novick, Sheldon M. 1996. *Henry James: The Young Master*. New York: Random House.

Nozick, Robert. 1981. *Philosophical Explanations*. Cambridge, Mass.: Harvard University Press.

Nussbaum, Martha C. 1997. *Cultivating Humanity: A Classical Defense of Reason in Liberal Education*. Cambridge, Mass.: Harvard University Press.

Oakeshott, Michael. 1989. *The Voice of Liberal Learning: Michael Oakeshott on Education*. Ed. Timothy Fuller. New Haven: Yale University Press.

– 1991. *Rationalism in Politics and Other Essays*. Indianapolis: Liberty Press.

O'Brien, Geoffrey. 2000. *The Browser's Ecstasy: A Meditation on Reading*. Washington, DC: Counterpoint.

Pascal, Blaise. 1995. *Pensées and Other Writings*. Trans. H. Levi. Oxford and New York: Oxford University Press.

Passmore, John. 1961. *Philosophical Reasoning*. London: Gerald Duckworth and Co.

– 1980. *The Philosophy of Teaching*. London: Gerald Duckworth and Co.

Pater, Walter. 1910. *Plato and Platonism*. London: Macmillan and Co.

Pattison, Mark. 1885. *Memoirs*. London: Oxford University Press.

– 1889. *Essays by the Late Mark Pattison*. Oxford: Oxford University Press. 2 vols.

Peckham, Morse. 1970. *Victorian Revolutionaries*. New York: George Braziller.

Pelikan, Jaroslav. 1992. *The Idea of the University: A Reexamination*. New Haven: Yale University Press.

Perkins, David. 1988. 'Taking Stock after Thirty Years.' *Teaching Literature. What Is Needed Now*. Ed. James Engell and David Perkins. Cambridge, Mass.: Harvard University Press, 111–17.

– 1992. *Is Literary History Possible?* Baltimore: Johns Hopkins University Press.

Pocklington, Tom, and Alan Tupper. 2002. *No Place to Learn: Why Universities Aren't Working*. Vancouver: University of British Columbia Press.

Polanyi, Michael. 1958. *Personal Knowledge: Towards a Post-Critical Philosophy*. Chicago: University of Chicago Press.

Pope, Alexander. 1963. *The Poems of Alexander Pope*. Ed. John Butt. New Haven: Yale University Press.

Postman, Neil, and Charles Weingartner. 1969. *Teaching as a Subversive Activity*. New York: Delacorte Press.

Price, Reynolds. 1999. *Letter to a Man in the Fire: Does God Exist and Does He Care?* New York: Scribner.

Pritchard, William H. 1984. *Frost: A Literary Life Reconsidered.* New York: Oxford University Press.

Rawls, John. 1971. *A Theory of Justice.* Cambridge, Mass.: Harvard University Press.

Raymond, W.O. 1950. *The Infinite Moment and Other Essays on Robert Browning.* Toronto: University of Toronto Press.

Readings, Bill. 1996. *The University in Ruins.* Cambridge, Mass.: Harvard University Press.

Redfield, James M. 1967. 'Platonic Education.' *The Knowledge Most Worth Having.* Ed. Wayne Booth. Chicago: University of Chicago Press.

Ricks, Christopher. 1987. *The Force of Poetry.* Oxford: Oxford University Press.

Rilke, Rainer Maria. 1992. *Letters to a Young Poet.* Trans. Joan M. Burnham. Novato, Calif.: New World Library.

Rorty, Richard. 1979. *Philosophy and the Mirror of Nature.* Princeton: Princeton University Press.

– 1999. *Philosophy and Social Hope.* London: Penguin.

Rothman, Ellen Lerner. 1999. *White Coat: Becoming a Doctor at Harvard Medical School.* Perennial: New York.

Royce, Josiah. 1901. *The World and the Individual.* 2 vols. New York: Macmillan.

Russell, Bertrand. 1963. 'The Negative Theory of Education.' *The Teacher and the Taught.* Ed. Ronald Gross. New York: Dell, 212–22.

Ryan, Alan. 2001. 'Schools: The Price of "Progress."' *New York Review of Books.* Vol. 48, no. 3, 18–21.

Sabin, Margery. 1997. 'Evolution and Revolution: Change in the Literary Humanities, 1968–1995.' *What's Happened to the Humanities?* Ed. Alvin Kernan. Princeton: Princeton University Press, 84–103.

Said, Edward W. 1999. *Out of Place: A Memoir.* New York: Alfred A. Knopf.

Sarton, May. 1961. *The Small Room.* New York: W.W. Norton and Co.

Schön, Donald A. 1990. *Educating the Reflective Practitioner: Toward a New Design for Teaching and Learning in the Professions.* San Francisco and Oxford: Jossey-Bass.

Schopenhauer, Arthur. 1970. *Essays and Aphorisms.* Trans. R.J. Hollingdale. Harmondsworth: Penguin.

Sewell, Elizabeth. 1952. *The Field of Nonsense.* London: Chatto and Windus.

Shattuck, Roger. 1999. *Candor and Perversion: Literature, Education, and the Arts.* New York: W.W. Norton and Co.

Shaw, George Bernard. 1946. *The Revolutionist's Handbook* from *Man and Superman.* London and Edinburgh: Penguin.

– 2002. *The Sayings of Bernard Shaw.* Ed. Joseph Spence. London: Duckworth.

Shelley, Percy Bysshe. 1951. *Selected Poetry and Prose of Percy Bysshe Shelley.* Ed. Carlos Baker. New York: Modern Library.

Shklar, Judith N. 1988. 'Why Teach Political Theory?' *Teaching Literature: What Is Needed Now.* Ed. James Engell and David Perkins. Cambridge, Mass.: Harvard University Press, 151–60.

Sparshott, F.E. 1972. *Looking for Philosophy.* Montreal and London: McGill-Queen's University Press.

– 1994. *Taking Life Seriously: A Study of the Argument of the Nicomachean Ethics.* Toronto: University of Toronto Press.

Spoto, Donald. 1998. *The Hidden Jesus: A New Life.* New York: St Martin's Press.

Steiner, George. 2003. *Lessons of the Masters.* Cambridge, Mass.: Harvard University Press.

Stevens, Wallace. 1974. *The Collected Poems of Wallace Stevens.* New York: Alfred A. Knopf.

Strauss, David Friedrich. 1846. *The Life of Jesus, Critically Examined.* Trans. George Eliot. 3 vols. London: Chapman.

Taleb, Nassim Nicholas. 2001. *Fooled by Randomness: The Hidden Role of Chance in the Markets and in Life.* New York: Texere.

Tennyson, Alfred. 1987. *The Poems of Tennyson.* Ed. Christopher Ricks. 3 vols. London: Longmans, Green.

Thoreau, Henry David. 1965. *Walden. American Literary Masters.* Ed. Charles R. Anderson. New York: Holt, Rinehart and Winston.

Traherne, Thomas. 1966. *Poems, Centuries, and Three Thanksgivings.* Ed. Anne Ridler. Oxford: Oxford University Press.

Turner, Frank. 2002. *John Henry Newman: The Challenge to Evangelical Religion.* New Haven: Yale University Press.

Unamuno, Miguel de. 1972. *The Tragic Sense of Life in Men and Nations.* Trans. Anthony Kerrigan. Princeton: Princeton University Press.

Vaugn, Henry. 1981. *The Complete Poems.* New Haven: Yale University Press.

Vendler, Helen. 1988a. 'What We Have Loved.' *Teaching Literature: What Is Needed Now* Ed. James Engell and David Perkins. Cambridge, Mass.: Harvard University Press, 13–25.

– 1988b. *The Music of What Happens: Poems, Poets, Critics.* Cambridge, Mass.: Harvard University Press.

Verene, Donald Phillip. 1997. *Philosophy and the Return to Self-Knowledge.* New Haven: Yale University Press.

Vidler, Alec. 1964. 'The Future of Divinity.' *Crisis in the Humanities.* Ed. J.H. Plumb. Harmondsworth: Penguin, 82–95.

Weinberg, Steven. 2001. 'Can Science Explain Everything? Anything?' *New York Review of Books.* Vol. 48, no. 9. 31 May, 47–50.

White, Morton G. 1982. *Religion, Politics, and the Higher Learning.* Westport, Conn.: Greenwood Press.

Whitehead, A.N. 1925. *Science and the Modern World.* New York: Macmillan.

– 1933. *Adventures of Ideas.* London: Collier-Macmillan.

– 1949. *The Aims of Education.* New York: Mentor.

Wilde, Oscar. 1950. *The Picture of Dorian Gray.* Harmondsworth: Penguin.

– 1968. *Literary Criticism of Oscar Wilde.* Ed. Stanley Weintraub. Lincoln: University of Nebraska Press.

Wittgenstein, Ludwig. 1961. *Tractatus Logico-Philosophicus.* Trans. D.F. Pears and B.F. McGuinness. London: Routledge and Kegan Paul.

– 1972. *Ludwig Wittgenstein. Philosophical Investigations.* Trans. G.E.M. Anscombe. Oxford: Basil Blackwell.

Wordsworth, William. 1940–9. *The Poetical Works of William Wordsworth.* Ed. Ernest de Selincourt and Helen Darbishire. 5 vols. Oxford: Clarendon Press.

– 1950. Preface to second edition of *Lyrical Ballads* in *William Wordsworth: Selected Poetry.* Ed. Mark Van Doren. New York: Modern Library.

Yeats, W.B. 1960. *The Collected Poems of W.B. Yeats.* New York: Macmillan.

Index

Numerals in italic type indicate the locations of the main discussions.

By W. David Shaw

Origins of the Monologue: The Hidden God

Alfred Lord Tennyson: The Poet in an Age of Theory

Elegy and Paradox: Testing the Conventions

Elegy and Silence: The Romantic Legacy

Victorians and Mystery: Crises of Representation

The Lucid Veil: Poetic Truth in the Victorian Age

Tennyson's Style

The Dialectical Temper: The Rhetorical Art of Robert Browning